THE GENESIS AND
CLASSIFICATION OF
COLD SOILS

THE GENESIS AND CLASSIFICATION OF COLD SOILS

Samuel Rieger

Retired Soil Scientist
Soil Conservation Service
U.S. Department of Agriculture
Palmer, Alaska

ACADEMIC PRESS   1983

A Subsidiary of Harcourt Brace Jovanovich, Publishers

New York   London
Paris   San Diego   San Francisco   São Paulo   Sydney   Tokyo   Toronto

ACADEMIC PRESS, INC.
111 Fifth Avenue, New York, New York 10003

United Kingdom Edition published by
ACADEMIC PRESS, INC. (LONDON) LTD.
24/28 Oval Road, London NW1 7DX

Library of Congress Cataloging in Publication Data

Rieger, Samuel, Date
  The genesis and classification of cold soils.

  Includes index.
  1. Soil formation. 2. Soils--Classification.
3. Cold regions. 4. Soil temperature. I. Title.
S592.2.R53 1983   553.6   83–2541
ISBN 0–12–588120–7

PRINTED IN THE UNITED STATES OF AMERICA

83 84 85 86   9 8 7 6 5 4 3 2 1

CONTENTS

Chapter 3

THE UNITED STATES SOIL TAXONOMY

Chapter 4

ENTISOLS

Chapter 5

SPODOSOLS

Chapter 6

ALFISOLS

Chapter 12

HISTOSOLS

Chapter 13

THE TAXONOMIES OF CANADA, THE U.S.S.R., AND THE FAO

PREFACE

Soils of cold regions have not commanded in the past as much attention from soil scientists as those of temperate or tropical areas because of their more limited suitability for agriculture. Interest in these soils has increased considerably in recent years, however, along with increased awareness of cold environments generally. The kinds of soils that exist in cold regions and the physical and chemical processes involved in their development are now reasonably well known. Most of the soil-forming processes are not unique to cold areas, but low temperatures, deep freezing, and, in much of the region, the existence of a perennially frozen substratum, modify them to produce soils with unique properties.

This book summarizes existing knowledge of the processes involved in the development of the principal kinds of soils that occur in cold regions and introduces readers who are not themselves pedologists to the classification of those soils. This is a subject, unfortunately, that can be rather confusing. In contrast with the situation in most disciplines that deal with natural objects, there is no generally accepted system of soil classification. Many countries with active soil survey organizations have adopted "official" taxonomies, but these differ in important ways—not least of which is the terminology used in naming and describing soils. Individual pedologists have devised still other classification schemes that emphasize different soil properties and have differing methods for grouping soils to make the distinctions they consider to be most useful. None of these has attained the stature and general acceptance of the Linnaean system for classifying living organisms.

As may be expected, this diversity has made communication among pedologists of different backgrounds quite difficult. To avoid confusion in a work of this kind, it is necessary to select one taxonomic scheme and to use its concepts and terminology exclusively in the description of soils, regardless of the origin of the information presented. The one selected for use here, because of my familiarity with it and my belief that it is—although not, as will seen, beyond criticism—the most satisfactory soil classification system yet devised, is the Soil Taxonomy of the United States Department of Agriculture. The salient features of that system are outlined in Chapter 3 and in the chapters devoted to soil orders, but no attempt is made to repeat the detailed specifications for diagnostic horizons and taxa that appear in *Soil Taxonomy* (U.S. Dept. Agr. Handbook 436, 1975). Ideally, that volume will be consulted frequently as the reader works through this one.

It is my hope that workers in other disciplines who have an interest in cold regions, as well as pedologists who lack familiarity with soils of the colder parts of the world, will find this book to be a useful introduction to the soil component of low-temperature ecological systems.

The thorough and helpful review of the manuscript by Professor Kaye R. Everett of the Institute of Polar Studies, Ohio State University, is gratefully acknowledged.

Samuel Rieger

Chapter *1*

TEMPERATURE
RELATIONSHIPS
IN COLD SOILS

Cold soils occupy roughly one-third of the world's land area. They occur, of course, in areas with cold climates, but within this obvious relationship are many variations in temperature and precipitation patterns that are reflected in soil temperature and moisture regimes and in the kinds of vegetation the soils support. Two major climatic types, the maritime type and the continental type, can be recognized, but each of these has varying degrees of expression, and climates intermediate between the two types are common. In maritime climates precipitation is high and is fairly uniformly distributed over the year, and temperature differences between winter and summer months are not great. Cold continental climates have relatively low annual precipitation, most of which occurs in a short warm summer, and have long cold winters. A third climatic type, in which the temperature is nearly constant throughout the year, is restricted to high mountains of low latitudes. A wide range in precipitation distribution and amounts exists within this mountain type.

I. Temperature Belts

Two broad temperature belts, each with its maritime and continental component, can be identified largely by their characteristic vegetation. The colder of these is the *polar and alpine* belt of high latitudes and altitudes. Here

1

temperatures are low all year, and even the warmest month has a mean air temperature lower than 10° C. No trees survive in this region. Dense and succulent tundra vegetation is most common in maritime climates and generally more shrubby vegetation in continental climates, although sedges are a major component in both areas.* The height and density of the vegetation in all areas is directly related to the mean summer temperature. The colder part of the polar and alpine belt, in which precipitation is very low and the mean air temperature of the warmest month is lower than 4° C, has very sparse and low-growing tundra vegetation that is confined to low points in the microrelief (Tedrow, 1968). The only land areas with this *high arctic,* or *polar desert,* climate are islands in the Arctic and Antarctic Oceans, high northern mountains, and the northernmost tip of the Taymyr Peninsula of Siberia. The coldest ice-free land areas of all are in Antarctica, where annual precipitation is less than 50 mm and temperatures are rarely higher than 0° C. These areas and others in high northern mountains where the mean July temperature is lower than 2° C are virtually free of higher vegetation.

The second, or *boreal belt,* includes the area between the southern or lower limit of the polar and alpine belt and, approximately, the boundary between the dominantly coniferous boreal forest, or taiga, and the mixed forest of mid latitudes. This warm limit of the boreal belt may be defined in various ways, but for the present purpose it can be taken to coincide with the boundary between cold soils, as defined later in this chapter, and temperate or mesic soils. The forests increase in density and in proportion of hardwoods from north to south and from upper to lower elevations in the mountains. In the northern or upper parts of the belt, forests on better drained soils alternate with tundra on poorly drained soils. With decreasing latitude or elevation the proportion of the area with tundra-like vegetation even on wet soils decreases and eventually is confined to peat bogs. Some areas of natural grasslands are also included in the boreal belt. These occur under continental climates with very low precipitation mostly (but not exclusively) in the warmer part of the belt and, generally in association with tundra-like vegetation, in areas with strongly maritime climates and low summer temperatures. Lack of adequate moisture is responsible for the existence of steppe rather than forest vegetation in the cold continental areas; summer temperatures that are too low for tree growth probably account for the absence of trees in most of the grassy maritime areas. In some areas, however, including much of Iceland and the Aleutian Islands, the mean temperature of the warmest month is higher than 10° C, the theoretical lower limit for tree survival. In Iceland the lack of trees is attributed to destruction of the original forest by early settlers (Johannesson, 1960), but in the Aleutians

---

* More precise formulations of the climatic parameters that define the warm limits of tundra vegetation are available (Barry and Ives, 1974), but discussion of them is beyond the scope of this volume.

the reason is not readily apparent. It seems likely that climatic factors other than temperature, such as the strong winds and frequent fog that affect the area, are responsible for the suppression of tree growth.

II. Soil Temperature Regimes in Cold Areas

It has long been recognized that systems of soil classification, if they are to be useful, must make taxonomic distinctions based on soil temperature. Any definition of soil temperature regimes must be arbitrary to some extent, since gradations between them are generally diffuse rather than abrupt, but each of the major classification systems now in use in cold regions attempts to define temperature regimes so that they reflect important differences in soil properties related to temperature, in the natural vegetation supported by the soils, or in existing or potential uses of the soils.

A. TEMPERATURE CLASSES IN THE UNITED STATES TAXONOMY

In the soil taxonomy adopted by the United States Department of Agriculture (Soil Survey Staff, 1975), two classes of cold soils are recognized. *Cryic* soils are defined as those with a mean annual *soil* temperature lower than 8° C but higher than 0° C and with relatively low mean summer (June through August in the northern hemisphere) soil temperatures at a depth of 50 cm. For well-drained soils under natural vegetation and protected by a layer of organic litter on the soil surface, the mean summer soil temperature must also be lower than 8° C. For cleared well-drained soils with the litter removed or mixed with the upper part of the mineral soil it must be lower than 15° C. For mineral soils with impeded internal drainage that are saturated for significant periods during the summer, the mean summer temperature limits are 6° C for undisturbed soils and 13° C for cleared soils. A different definition applies to some organic soils associated with cryic mineral soils; this is discussed in Chapter 12. Cryic soils occur in the boreal climatic belt and in alpine areas at lower latitudes.

Pergelic soils are those with a mean annual soil temperature of 0° C or lower and therefore with a layer at some depth in which the temperature is perennially at or below 0° C. Most of these soils are underlain by solid ice-rich permafrost, but in some coarse-textured soils there is no ice even at temperatures below 0° C because of a lack of moisture; this condition is known as dry permafrost. Most, but not all, soils in the polar and alpine belt are pergelic soils. Many forested soils in the colder part of the boreal belt, especially under continental climates, also have pergelic temperature regimes.

In the northern hemisphere cryic and pergelic soils occupy all of Alaska, most of Canada exclusive of a relatively narrow strip along its southern border, the North Atlantic islands including Greenland and Iceland, the northernmost portions of the British Isles and Scandinavia, northern European U.S.S.R., virtually all of Siberia, Mongolia, and the northern provinces of China, and high mountains south of those areas. In the southern hemisphere, apart from Antarctica, these soils occur principally in mountains and in the southernmost part of South America (Mikhaylov, 1970). The two zones are not mutually exclusive. Patches of pergelic soils occur commonly in the area dominated by cryic soils, especially in areas with continental climates; the proportion of soils with permafrost decreases with decreasing latitude or elevation. To a lesser extent patches of cryic soils occur in forested regions dominated by pergelic soils. In high mountains at lower latitudes, pergelic and cryic soils occur generally in areas above tree line and, in some places, in forested subalpine areas (Johnson and Cline, 1965; Achuff and Coen, 1980).

B. TEMPERATURE CLASSES IN CANADIAN CLASSIFICATION

In the Canadian system of soil classification (Canada Soil Survey Committee, 1978), three classes rather than two are used to characterize cold soils. *Arctic,* or extremely cold, soils have mean annual temperatures lower than -7° C and mean summer temperatures lower than 5° C at a depth of 50 cm. Soils of this class are in the high arctic or polar desert portion of the polar temperature belt; all of them have perennially frozen substrata. *Subarctic,* or very cold, soils have mean annual temperatures of -7°-2° C and mean summer temperatures of 5°-8° C. They occur both in the polar and alpine belt under tundra vegetation and in the boreal belt under forest. Most of these soils are underlain by permafrost, but some poorly drained soils in coastal areas and some well-drained soils of floodplains are not. In general, however, the arctic and subarctic temperature classes together correspond with the pergelic class in the United States Soil Taxonomy. *Cryoboreal,* or cold to moderately cold, soils are defined as having annual temperatures of 2°-8° C and mean summer temperatures of 8°-15° C. These soils do not have perennially frozen substrata. The cryoboreal class is roughly equivalent to the cryic temperature class in the United States Taxonomy despite the discrepancy in summer soil temperature limits for uncleared soils.

C. TEMPERATURE RELATIONSHIPS IN U.S.S.R. CLASSIFICATION

The official soil classification system of the U.S.S.R. (Rozov and Ivanova, 1967a, 1967b) does not provide for separate temperature classes of soils, but temperature is one of the factors considered in defining ''ecological-genetic''

classes that are basic features of the taxonomy. Virtually all cold soils, as defined in the United States and Canadian systems, are included in three of these
classes. The class, *soils of tundra and arctic regions,* includes all soils under tundra
vegetation. *Soils of frozen taiga regions* includes most soils of the boreal belt that
are underlain by permafrost. These two classes correspond with the pergelic
temperature class in the United States Taxonomy. *Soils of taiga–forest boreal
regions* includes soils of the boreal belt that have no permafrost, but extends
somewhat beyond the warm limit of the cryic class of the United States Taxonomy and the cryoboreal class of the Canadian system. Despite their names,
the latter two classes in the U.S.S.R. classification also include some soils that
support steppe rather than forest vegetation.

D. HEAT UNITS

In the U.S.S.R. and Canada the temperature classes are related to heat
units defined as degrees or degree days above a specified temperature. The sum
of annual surface temperatures above 10° C (the sum of the mean daily degrees
above 10° C in the course of a year) is used as part of the definition of
ecological-genetic classes in the U.S.S.R. *Soils of tundra and arctic regions* occur
in areas with less than 600°. This value is also used in the U.S.S.R. to define
the polar belt (Kovda *et al.,* 1967). *Soils of frozen taiga regions* are confined to areas
with 600°–800°, but *Soils of taiga–forest boreal regions* may occur anywhere in the
boreal belt, which is defined as having 600°–2400° above 10° C.

In Canada the number of days with temperatures of 5° C or higher (the
growing season) and the number of daily degrees above 5° C during the growing season are used as secondary criteria to characterize soil temperature
classes. The nominal growing season for the arctic class is less than 15 days. For
the subarctic class it is shorter than 120 days, and there are less than 555
degree-days above 5° C during the growing season. The cryoboreal class normally has a growing season of 120–220 days and 555–1250 degree-days.

No consideration is given to heat units in the United States Taxonomy
because, although air temperature is obviously the principal determinant of soil
temperatures, extremes of summer temperature that may influence the
number of degree days are not always reflected in either mean annual or mean
summer soil temperatures. Many other factors, not directly related to air
temperature, also influence the temperature of the soil. Among these are the
kind and density of the vegetation cover and the thickness and density of
the organic litter on the soil surface, the direction and steepness of slope, the
thickness and character of the snow cover in winter, the texture of the soil, the
moisture content of the soil, and characteristics of the climate itself.

III. Factors Affecting Soil Temperatures

A. RELATIONSHIPS BETWEEN SOIL AND AIR TEMPERATURES

Soils in areas with continental climates generally are much colder and freeze more rapidly to greater depths than soils under maritime climates. In forested soils under the maritime climate of northwestern U.S.S.R., for example, the temperature at a depth of 20 cm is rarely lower than -3° C, and maximum depth of freezing does not exceed 1 m, whereas under the cold continental climate of southeastern Siberia the temperature of the upper horizons of forested soils is -20° C or lower and the soils freeze to depths of several meters (Sokolov and Sokolova, 1962). At the southern edge of the forest in European U.S.S.R., noncryic cultivated soils freeze only to depths of 50–70 cm and remain frozen only 4–5 months, but at the edge of the forest in south central Siberia, soils freeze to depths of almost 3 m and are frozen in some part for 9–10 months (Veredchenko, 1959).

In mid latitudes the mean annual temperature of most well-drained soils can be estimated fairly accurately by adding 1° C to the mean annual air temperature (Soil Survey Staff, 1975). This relationship holds fairly well in boreal maritime climates such as that at Mustiala, Finland (Chang, 1958), where the mean soil temperature is less than 2° C higher than the mean air temperature, but in other cold climates it may not. At Barrow, Alaska, the mean soil temperature at a depth of only 16 cm in a well-drained, gravelly soil was found to be -8.2° C, as opposed to a mean annual air temperature of about -13° C (Kelley and Weaver, 1969). At Edmonton, Alberta, the mean soil temperature is about 3.6° C higher than the mean air temperature (Toogood, 1976). In the northeastern part of European U.S.S.R., the mean soil temperature is more than 3.5° C higher; in western Siberia under a continental climate, it is 5° C higher; and in the strongly continental Amur region of southeastern Siberia it is as much as 6.2° C higher (Golovin, 1962). Well-drained soils with no permafrost occur commonly in areas with mean annual air temperatures below -4°C (Brown, 1960). In general, the difference between the two means tends to be greater at higher latitudes and in more strongly continental climates.

Most of the difference between mean annual soil and air temperature is the result of strong winter temperature deviations. In the Amur region, for example, mean summer soil temperatures at a depth of 20 cm under grass are less than 1° C lower than mean summer air temperatures, but in winter, even under a thin snow cover, mean soil temperatures at that depth are 7°–13° C warmer than mean air temperatures (Golovin, 1962). In most cold areas the principal reason for retention of heat in the soil despite low winter air temperatures is the insulating cover of snow that commonly reaches its maximum thickness during the coldest part of the winter, but also important, especially in cold continental climates with little snow, is the latent heat of soil water that is released during freezing.

B. EFFECT OF VEGETATIVE COVER

The character of the vegetative cover and its associated mat of organic litter on the soil surface has a significant effect on both summer and mean annual soil temperatures. A dense forest canopy shades the soil in summer and thus retards warming. This results in lower mean summer soil temperatures in heavily forested soils than in grass-covered or sparsely forested soils. At high elevations in western Montana, a forested, well-drained soil had a mean summer temperature at a depth of 50 cm about 4.7° C lower than the mean summer air temperature. In a nearby well-drained soil under grass, the mean summer soil temperature was only 1.6° C lower than the air temperature, and in a transitional area between the two kinds of vegetation, it was 3.9° C lower (Munn *et al.*, 1978)(Fig. 1). The rate of cooling in winter was much less affected by the vegetation, probably because of a uniformly thick snow cover over the entire region, and winter soil temperatures were nearly the same at all sites. Overall, heavily forested soils in the area have mean annual soil temperatures lower than air temperatures, whereas the mean annual temperatures of grass-covered soils are higher than that of the air (Mueller, 1970).

Under a forest cover and its associated ground litter, the seasonal fluctuation of soil temperature is much lower than in open areas, especially in the upper layers of the soil. In the strongly continental Amur region, temperatures in the upper 30 cm are 5° C warmer in cleared fields than under forest in midsummer, but at depths between 1 and 3 m, the forested soil is warmer than the cleared soil throughout the year (Pustovoytov, 1964). The annual amplitude of temperature in upper horizons is 10°–13° C greater in cleared fields than under forest (Pustovoytov, 1962).

Shading by the forest canopy can have a large effect on summer soil temperatures and can to some extent counteract the temperature differences that can normally be expected between soils at different latitudes. In northern European U.S.S.R. the highest temperature reached at a depth of 20 cm in the course of the summer is generally only 16° C under a dense coniferous–broadleaf forest near Moscow (lat 55°45′ N), whereas at Rybinsk (lat 58°5′ N) under a less complete pine canopy and at Arkhangelsk (lat 64°40′ N) under sparse pine and brush, it is 14°–15° C. The more southerly soils warm to greater depth, however, and have longer warm periods (Yelagin and Izotov, 1968).

Much of the effect of forest on soil temperature is indirect in that the snow accumulation in winter is soft and dry and does not blow away. In the tundra,

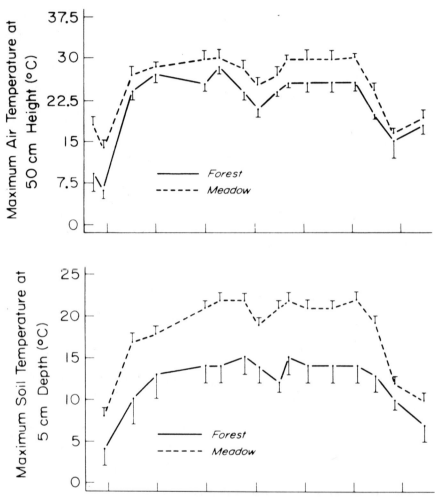

FIGURE 1. *Maximum and minimum air temperatures and soil temperatures at two depths under forest and meadow vegetation in the Rocky Mountains of Montana during the summer of 1970 (Munn et al., 1978). Standard errors are shown for each sampling data. (Reprinted from* Soil Science Society of America Journal, *Volume 42, page 983, 1978. By permission of the Soil Science Society of America.)*

by contrast, the snow is normally drifted and hard-packed and is therefore a less efficient insulator (Beckel, 1957), or it may be completely blown away. Drifting and packing of snow is also common on grasslands and unprotected cleared fields. At Krasnoyarsk in south central Siberia the average depth of freezing under steppe vegetation is 280 cm, and the mean annual soil temperature is 2.7° C, but under forest the depth of freezing is only 115 cm, and the mean soil temperature is 4.7° C. The difference is attributable to the retained snow cover on the forested soil (Veredchenko, 1959).

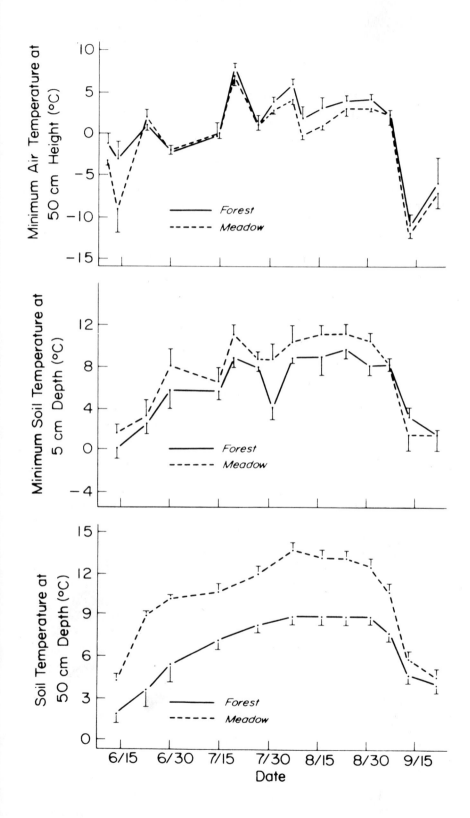

In soils underlain by permafrost, the depth of thaw in summer is related to the height and density of the vegetation and the thickness of the organic mat below the living plants. In dense shrubby tundra above tree line in interior Alaska, removal of the vegetation while leaving intact the thick underlying mat of darker decaying organic matter resulted in increased depth of thaw two-thirds as great as that resulting from removal of both the vegetation and the organic mat (Brown *et al.*, 1969). Similarly, removal of trees from a forested soil with permafrost near Fairbanks without disturbing the forest litter increased the depth of thaw from 1 m to more than 3 m (Kallio and Rieger, 1969). Under cultivation the depth of thaw increased still further (Fig. 2). Forest fires have a comparable effect on the permafrost table and, in hummocky soils, may result in temporary changes in the height and shape of the hummocks (Pettapiece, 1974) (Fig. 3).

In tundra polygons with barren "frost circles," thawing in the spring begins as much as 1 month earlier under the frost circles than under vegetated portions of the polygons, and both the maximum depth of thaw and attained soil temperatures during the summer are greater under the frost circles (Vasil'yev-skaya, 1979).

The effect of perennial grass vegetation on soil temperatures is similar to that of tundra but is not as pronounced. In areas with cool maritime climates, a layer of decaying residues that may attain a thickness of 20 cm or more accumulates under grass stands and affects soil temperatures.

C. EFFECT OF ORGANIC MAT

In general, dense vegetation is associated with a thick organic mat. Together, they insulate the soil and the underlying permafrost from both seasonal and diurnal temperature fluctuations, with the degree of protection related directly to the thickness of the organic cover. If thick enough it can keep soils frozen even in areas where the mean annual temperature is somewhat higher than 0° C. Poorly drained, loamy soils with tundra vegetation and an organic mat 40 cm thick in southwestern Alaska, where the mean annual air temperature is about .5° C, thaw only to the base of the mat during the summer. The depth of thaw is no greater than that in northern Alaska, where the mean annual temperature is lower than -12° C, but where the vegetation on similar soils is shorter and the organic mat is only 5 cm thick (Rieger, 1974). Growth of vegetation and development of an organic mat on recent alluvial deposits can eventually result in the formation of a perennially frozen substratum where none existed before (Viereck, 1970).

The organic cover is also effective in damping diurnal temperature variations. At Barrow, Alaska, the temperature range at the surface of a well-drained, sandy soil with 8 cm of organic litter was nearly 12° C on a clear sum-

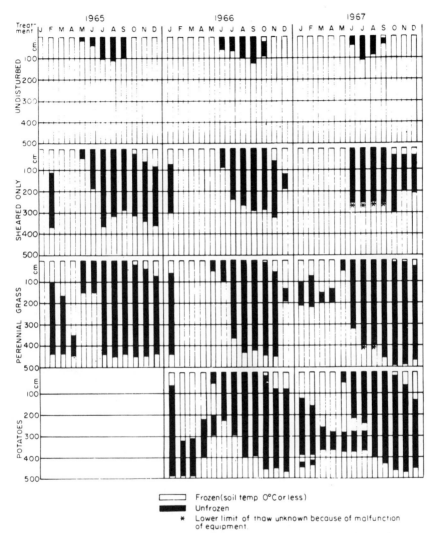

FIGURE 2. *Mean monthly frozen and unfrozen zones under native black spruce forest, sheared conditions (removal of forest and all but about 2 cm of forest litter), perennial grass, and potatoes over a 3-year period in a soil with permafrost in interior Alaska (Kallio and Rieger, 1969). (Reprinted from* Soil Science Society of America Proceedings, *Volume 33, page 430, 1969. By permission of the Soil Science Society of America.)*

mer day. At a depth of 4 cm below the mat, it was 3.5° C, and at 16 cm, only 1.4° C (Kelley and Weaver, 1969) (Fig. 4). In winter the insulating effect of the organic mat is considerably reduced. The thermal conductivity of frozen organic matter is about four times higher than that of unfrozen peat with the same moisture content (Zoltai and Tarnocai, 1971). Thus, the same organic

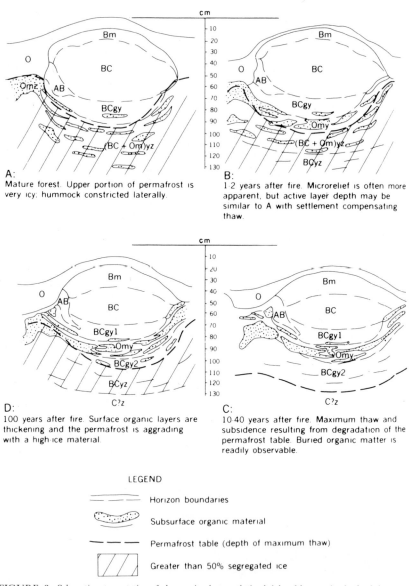

A:
Mature forest. Upper portion of permafrost is very icy; hummock constricted laterally.

B:
1-2 years after fire. Microrelief is often more apparent, but active layer depth may be similar to A with settlement compensating thaw.

D:
100 years after fire. Surface organic layers are thickening and the permafrost is aggrading with a high-ice material.

C:
10-40 years after fire. Maximum thaw and subsidence resulting from degradation of the permafrost table. Buried organic matter is readily observable.

LEGEND

══════════ Horizon boundaries

⬭⬭⬭ Subsurface organic material

─ ─ ─ Permafrost table (depth of maximum thaw)

▨ Greater than 50% segregated ice

FIGURE 3. *Schematic representation of changes in shape, relative height of hummock, depth of thaw, and organic matter distribution in a hummocky soil near Inuvik, Northwest Territories, Canada. Symbols represent horizon designations according to the Canadian system: Upper case letter designations are like those of the United States Taxonomy; ''g'' indicates gleying; ''m'' indicates other changes in color or structure or both from parent material; ''y'' indicates horizon affected by cryoturbation; and ''z'' indicates frozen condition. (Reprinted from Pettapiece [1974] by permission of the Agricultural Institute of Canada.)*

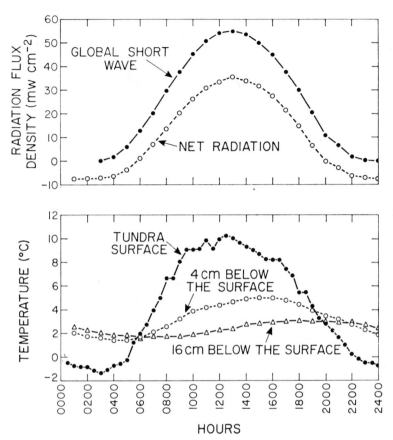

FIGURE 4. *Diurnal course of solar radiation and soil temperatures in a tundra soil near Barrow, Alaska, on a clear summer day. (From Kelley and Weaver [1969] by permission of the Arctic Institute of North America.)*

mat that in summer is an excellent insulating material, especially when it is partially dry, is in winter a fairly good thermal conductor. In the winter, however, it is normally covered with snow, which does have good insulating qualities and which offsets, at least in part, the ineffectiveness of the organic surface cover in protecting the soil against low air temperatures. In areas with continental boreal climates, summers are warm and most grass residues on well-drained soils decay rapidly or are incorporated with the underlying mineral soil. It has been observed, however, that in soils of these areas in which the permafrost table has receded to considerable depth after clearing for agriculture, it becomes shallow again if the field is kept in perennial grass (Boyd and Boyd, 1971).

D. EFFECT OF SLOPE STEEPNESS AND ASPECT

Major deviations of soil temperature from air temperature may be caused by direction and steepness of slope, especially at moderately high latitudes in areas with a strongly continental climate. At these latitudes the angle of the sun to the earth's plane is low. As a result, steep south-facing slopes (in the northern hemisphere) receive more direct solar radiation per unit area in both summer and winter than level areas at the same latitude and much more than north-facing slopes. In effect, a south-facing slope receives as much radiation as does level land at a lower latitude, and a north-facing slope as much as level land at a higher latitude. The "equivalent latitude" of a sloping soil may be calculated by the following formula (Lee, 1962):

$$\theta' = \arcsin\,(\sin k \cos h \cos \theta + \cos k \sin \theta)$$

where θ' is the equivalent latitude; k = degrees of slope from the horizontal; h = azimuthal degrees from north; and θ is the actual latitude. Thus, a 15° (approximately 27%) south-facing slope at lat 64° N has an equivalent latitude of about 49° N, and a 15° north-facing slope has an equivalent latitude of about 79° N. This does not mean that soils with these slope gradients necessarily have the characteristics of soils that commonly occur at the equivalent latitudes since factors other than potential solar radiation are also involved in the development of soil temperature regimes, but several striking examples of the effect of slope and aspect have been reported. In the vicinity of Fairbanks, Alaska, soils of steep, south-facing slopes are well-drained, have no permafrost, and support a productive white spruce–birch–aspen forest, whereas soils of equally steep, north-facing slopes are poorly drained, have ice-rich permafrost at shallow depths, and support vegetation dominated by Sphagnum moss and spindly, slow-growing black spruce (Krause *et al.*, 1959) (Fig. 5). In the tundra of northwestern Alaska, a very steep south-facing slope is grass-covered and has soil profile features resembling those of temperate grassland regions (Holowayachuk *et al.*, 1966). In strongly continental regions such as the Yana–Oymyakon Plateau of northeastern Siberia, soils of steep south-facing slopes support a xerophytic vegetation dominated by grasses and sagebrush rather than the forest characteristic of the region and are similar in many ways to dry steppe soils of southern Siberia (Naumov and Andreyeva, 1963; Volkovintser, 1974).

E. EFFECT OF SNOW COVER

Snow is an effective insulating material, especially when it is dry and loosely packed. It may be more important than the vegetative cover in reducing the depth of frost penetration in winter. In soils with permafrost, a heavy snow in early winter may prevent complete freezing of the soil above the permafrost table (Beckel, 1957). A snow cover thicker than 50 cm in the Hudson Bay

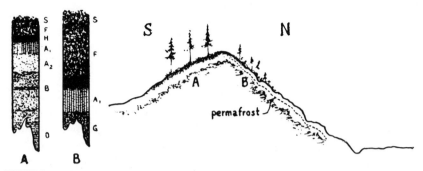

FIGURE 5. *Vegetation and soil on adjacent south-facing (A) and north-facing (B) slopes near Fairbanks, Alaska. (From Krause, Rieger, and Wilde [1959].)*

region gives the soil almost complete protection against fluctuations in air temperature (Zoltai and Tarnocai, 1971). In the Rocky Mountains of Montana, the soil temperature in late winter and early spring is nearly constant under a thick snow cover, regardless of aspect (Mueller, 1970) or vegetative cover (Munn *et al.*, 1978). Winter soil temperatures are much lower where the snow has been blown off, as on ridge tops, than where it accumulates in drifts. Where the pattern of drifting repeats year after year the thickness of the snow cover during the winter may be the critical factor in determining whether or not a perennially frozen substratum develops (Mackay and MacKay, 1974; Nicholson and Moore, 1977). A thick snow cover that typically accumulates only in late winter, after soil temperatures have dropped below freezing, and persists throughout the spring, as on north slopes in the Canadian Rockies, favors the development of permafrost (Achuff and Coen, 1980).

F. EFFECT OF SOIL TEXTURE AND ORGANIC MATTER

The retention and diffusion of heat in cold soils is affected by its particle-size distribution (texture) and its organic matter content. Coarse-textured soils generally transmit heat more rapidly and are more subject to variations in temperature than are fine-grained soils. Organic soils transmit heat most slowly. A mineral soil that has a high organic matter content and a dark surface color freezes more slowly and thaws more rapidly than a light-colored soil (Beckel, 1957). Gravelly soils with no appreciable organic matter accumulation at the surface in the polar desert were found to be warmer than would normally be expected for the latitude (French, 1970). Sandy and gravelly soils under tundra vegetation thaw to greater depths in the summer than fine-grained soils, under both poorly drained and well-drained conditions (Rieger, 1974). In southeastern Siberia mean summer temperatures of sandy soils are higher than mean air temperatures, but those of clayey and peaty soils are cooler (Golovin,

1962). The mean annual temperatures in all of these soils are about the same (Pustovoytov, 1964).

G. EFFECT OF SOIL MOISTURE

As a rule, under the same climatic conditions, poorly drained soils freeze less deeply than well-drained soils. In northern European U.S.S.R., for example, the depth of freezing is 15–25 cm greater in well-drained than in poorly drained forested soils (Izotov, 1968). Poorly drained soils also have, as is recognized in the definition of cryic soils in the United States Taxonomy, lower mean summer temperatures than well-drained soils. The presence of small quantities of moisture, however, does not always result in lower summer soil temperatures. Because the thermal conductivity of water is higher than that of air, soils that have large air spaces between particles—sand and gravel, primarily—warm and cool more rapidly under normal moist field conditions than when they are absolutely dry. Similarly, wet peat is a less effective insulating material than dry peat (Beckel, 1957).

Although saturated soils warm more slowly than moist or dry mineral soils, heat may be conducted to greater depths where free water exists throughout the profile than where there is an air-water interface. In the Hudson Bay area, a sandy soil with permafrost that was saturated only below 60 cm thawed to a total depth of 234 cm, whereas a texturally identical soil with standing water on the surface thawed to 335 cm (Beckel, 1957). In Spitsbergen soils under standing water in depressions thawed to depths of 60–80 cm, compared with 15 cm in slightly elevated soils with a thick moss cover (Smith, 1956).

H. THE "ZERO CURTAIN" EFFECT

During freezing all of the soil below the freezing front remains at or above 0° C until the soil above the front is completely frozen (Kelley and Weaver, 1969). Accompanying this "zero curtain" effect is a steady migration of soil moisture to the freezing front, especially in fine-grained and poorly drained soils. This additional moisture further retards the advance of the freezing front into the soil. The net effect is that, from early winter on, the soil cools more slowly than the air, and the rate and extent of soil temperature reduction decreases with depth. In the spring and summer, the process is reversed, but except where there is a dense forest canopy or a thick mantle of organic matter on the soil surface, the equilibrium between soil and air temperature is attained more quickly. In the Amur region the mean spring thaw rate was found to be one and a half times faster than the mean freezing rate (Pustovoytov, 1962). This occurs partly because warming of the soils is primarily in response to direct solar radiation, whereas cooling depends on the slower diffusion of energy as a function of the soil–air temperature gradient (Munn *et al.*, 1978), and partly because the thermal conductivity of frozen soil is greater than that of either wet or dry unfrozen soil. As the soil becomes more deeply thawed the rate of increase in temperature slows and eventually becomes insignificant (Weedfall, 1963; Fedorova, 1974). In soils with permafrost depth to the frozen underlying layer changes little after midsummer.

I. COMBINED EFFECTS

It is evident that the temperature regime of a soil is the result of the interaction of several factors, both external and internal. These may act in concert to produce a regime that is considered ''normal'' for a region; this occurs most commonly in loamy soils with good surface drainage. In many soils, however, the influence of one or more of the factors may result in considerable modification of the ''normal'' temperature regime. For example, windblown areas such as ridge tops and treeless lowlands are colder than side slopes in winter because of the absence of a thick snow cover, but in summer the ridge tops as well as the side slopes are warmer than the wet lowlands. As a result, the highest mean annual temperatures occur on slopes, and the lowest in the low-lying areas (Mackay and MacKay, 1974). In hilly and mountainous areas in the boreal belt, especially, the varying effects of the temperature-modifying factors commonly result in a soil pattern that includes permafrost-free soils on southerly slopes, poorly drained soils with shallow permafrost in lowlands and on north-facing slopes, and soils with deeper permafrost on well-drained ridge tops. In this situation it is not possible to identify any single soil temperature regime that fully reflects the climatic characteristics of the area. Two or more regimes must be recognized as equally worthy of such recognition (Sokolov and Sokolova, 1962; Vishnevskaya, 1965).

IV. Temperatures in Relation to Soil Depth

Measurements of soil temperature at various depths throughout the year in southeastern and central Siberia (Veredchenko, 1959; Golovin, 1962; Pustovoytov, 1964), northwestern Canada (Mackay and MacKay, 1974), Alberta (Toogood, 1976), the Rocky Mountains of Montana (Mueller, 1970), and elsewhere (Soil Survey Staff, 1975) demonstrate that in any soil, the mean annual temperature is essentially the same at all depths although the amplitude of temperature—the difference between summer and winter temperatures—decreases with depth. At some depth in the soil, the temperature remains constant throughout the year (Fig. 6). In mid latitudes this depth generally is greater than 14 m where the water table is very deep and about 9 m

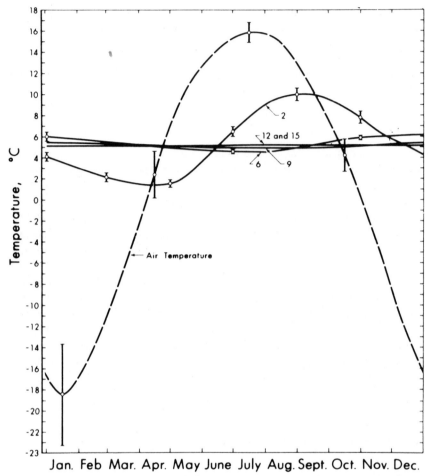

FIGURE 6. *Mean and standard deviations of soil temperatures at 2, 6, 9, 12, and 15 m, measured weekly, and air temperatures, measured daily, at Edmonton, Alberta, 1965–1975. (Reprinted from Toogood [1976] by permission of the Agricultural Institute of Canada.)*

where groundwater is present above that depth (Soil Survey Staff, 1975). In cold regions the depth to the layer of constant temperature ranges from 15 m in the delta of the Mackenzie River on the arctic coast of Canada (Mackay and MacKay, 1974) to only 3.2 m in mountain valleys of southeastern Siberia (Pustovoytov, 1964).

Temperature changes do not occur uniformly throughout the soil column. Instead, cooling or warming at the soil surface in response to changes in air temperature or direct solar radiation are reflected at depth only after a period of time. With increasing depth the time lag between the surface temperature change and the change in soil temperature increases, and the magnitude of change decreases. Diurnal or other surface temperature fluctuations of short duration seldom affect soil temperatures below a depth of about 50 cm. At greater depths, to the level of constant temperature, soil temperature change is in response only to the seasonal march of air temperatures. In cold soils, particularly, the progress of the cooling or heating wave in the soil may be quite slow. Cooling at a depth of several meters may continue for months after the soil at the surface has started to warm in the spring and, conversely, warming at depth may continue well after the onset of freezing at the surface (Pustovoytov, 1964; Kallio and Rieger, 1969). A frequently observed consequence of this phenomenon is the freezing of buried water pipes in the late winter or spring, well past the coldest part of the winter. The lag in temperature change at depth is due primarily to phase transformations of water during freezing and thawing. This is discussed in some detail in Chapter 2.

References

Achuff, P. L., and Coen, G. M. (1980). Subalpine cryosolic soils in Banff and Jasper National Parks. *Can. J. Soil Sci.* **60**, 579–581.

Barry, R. G., and Ives, J. D. (1974). Introduction. *In* "Arctic and Alpine Environments" (J. D. Ives and R. G. Barry, eds.), pp. 1–13. Methuen, London.

Beckel, D. K. B. (1957). Studies on seasonal changes in the temperature gradient of the active layer of soil at Fort Churchill, Manitoba. *Arctic* **10**, 151–183.

Boyd, W. L., and Boyd, J. W. (1971). Studies of soil microorganisms, Inuvik, Northwest Territories. *Arctic* **24**, 162–176.

Brown, J., Rickard, R., and Vietor, D. (1969). "The Effect of Disturbance on Permafrost Terrain." Cold Reg. Res. Eng. Lab., Spec. Rep. 138. U. S. Army, CRREL, Hanover, New Hampshire.

Brown, R. J. E. (1960). The distribution of permafrost and its relation to air temperature in Canada and the U. S. S. R. *Arctic* **13**, 163–177.

Canada Soil Survey Committee, Subcommittee on Soil Classification. (1978). "The Canadian System of Soil Classification." Can. Dept. Agr. Pub. 1646. Supply and Services Canada, Ottawa.

Chang, J. (1958). "Ground Temperature, II." Blue Hill Meteorol. Observ., Harvard U., Milton, Massachusetts.

Fedorova, N. M. (1974). Thermal and moisture regimes in a soil profile affected by prolonged seasonal freezing (middle taiga subzone, West Siberia). *Geoderma* **12**, 111–119.

French, H. M. (1970). Soil temperatures in the active layer, Beaufort Plain. *Arctic* **23**, 229–239.

Golovin, V. V. (1962). Description of the temperature regime of soils in the Amur region. *Sov. Soil Sci.* 213–217.

Holowaychuk, N., Petro, J. H., Finney, H. R., Farnham, R. S., and Gersper, P. L. (1966). Soils of Ogotoruk Creek watershed. *In* "Environment of the Cape Thompson Region, Alaska" (N. J. Wilimovsky, ed.), pp. 221–273. U. S. Atomic Energy Comm., Div. Tech. Inf. Ext., Oak Ridge, Tennessee.

Izotov, V. F. (1968). Pattern of soil freezing and thawing in the water-logged forests of the northern taiga subzone. *Sov. Soil Sci.,* 807–813.

Johannesson, B. (1960). "The Soils of Iceland." Univ. Res. Inst., Dept. Agr. Rep. B-13. Reykjavik.

Johnson, D. D., and Cline, A. J. (1965). Colorado mountain soils. *Adv. Agron.* **17,** 233–281.

Kallio, A., and Rieger, S. (1969). Recession of permafrost in a cultivated soil of interior Alaska. *Soil Sci. Soc. Am. Proc.* **33,** 430–32.

Kelley, J. J., and Weaver, D. F. (1969). Physical processes at the surface of the arctic tundra. *Arctic* **22,** 425–437.

Kovda, V. A., Lobova, Ye. V., and Rozanov, B. G. (1967). Classification of the world's soils—general considerations. *Sov. Soil Sci.* 427–441.

Krause, H. H., Rieger, S., and Wilde, S. A. (1959). Soils and forest growth on different aspects in the Tanana watershed of interior Alaska. *Ecology* **40,** 492–495.

Lee, R. (1962). Theory of the equivalent slope. *Mon. Weather Rev.* **90,** 165–166.

Mackay, J. R., and MacKay, D. K. (1974). Snow cover and ground temperatures, Garry Island, N. W. T. *Arctic* **27,** 287–296.

Mikhaylov, I. S. (1970). Main features of Chilean soils. *Sov. Soil Sci.* **2,** 1–7.

Mueller, O. P. (1970). Soil temperature regimes in a forested area of the northern Rockies. *Soil Sci.* **109,** 40–47.

Munn, L. C., Buchanan, B. A., and Nielsen, G. A. (1978). Soil temperatures in adjacent high elevation forests and meadows of Montana. *Soil Sci. Soc. Am. Jour.* **42,** 982–983.

Naumov, Ye. M., and Andreyeva, A. A. (1963). Soils of the Yana-Indigir upland slopes under steppe vegetation. *Sov. Soil Sci.,* 249–255.

Nicholson, H. M., and Moore, T. R. (1977). Pedogenesis in a subarctic iron-rich environment, Schefferville, Quebec. *Can. J. Soil Sci.* **57,** 35–45.

Pettapiece, W. W. (1974). A hummocky permafrost soil from the subarctic of northwestern Canada and some influences of fire. *Can. J. Soil Sci.* **54,** 343–355.

Pustovoytov, N. D. (1962). Influence of seasonal freezing on the water regime of Amur soils. *Sov. Soil Sci.,* 575–583.

Pustovoytov, N. D. (1964). Soil temperature regime in the plains of Amur River region. *Sov. Soil Sci.,* 352–359.

Rieger, S. (1974). Arctic soils. *In* "Arctic and Alpine Environments" (J. D. Ives and R. G. Barry, eds.), pp. 749–769. Methuen, London.

Rozov, N. N., and Ivanova, Ye. N. (1967a). Classification of the soils of the U. S. S. R. (Principles and a systematic list of soil groups). *Sov. Soil Sci.,* 147–155.

Rozov, N. N., and Ivanova, Ye. N. (1967b). Classification of the soils of the U. S. S. R. (Genetic description and identification of the principal subdivisions). *Sov. Soil Sci.,* 288–299.

Smith, J. (1956). Some moving soils in Spitsbergen. *J. Soil Sci.* **7,** 10–21.

Soil Survey Staff (1975). "Soil Taxonomy." U. S. Dept. Agr. Handbook 436, Washington, D.C.

Sokolov, I. A., and Sokolova, T. A. (1962). Zonal soil groups in permafrost regions. *Sov. Soil Sci.,* 1130–1136.

Tedrow, J. C. F. (1968). Pedogenic gradients of the polar regions. *J. Soil Sci.* **19,** 197–204.

Toogood, J. A. (1976). Deep soil temperatures at Edmonton. *Can. J. Soil Sci.* **56,** 505–506.

Vasil'yevskaya, V. D. (1979). Genetic characteristics of soils in a spotty tundra. *Sov. Soil Sci.* **11,** 390–401.

Veredchenko, Yu. P. (1959). Characteristics of the physical properties, water, and temperature of Krasnoyarsk forest-steppe soils. *Sov. Soil Sci.,* 1031–1040.

Viereck. L. A. (1970). Forest succession and soil development adjacent to the Chena River in interior Alaska. *Arc. Alp. Res.* **2,** 1–26.

Vishnevskaya, I. V. (1965). Mountain-taiga soils of the high taiga in eastern Tuva. *Sov. Soil Sci.,* 903–910.

Volkovintser, V. I. (1974). Soils of the dry steppe of the Yana-Oymyakon upland. *Sov. Soil Sci.* **6,** 142–150.

Weedfall, R. O. (1963). Variation of soil temperatures in Ogotoruk Valley, Alaska. *Arctic* **16,** 181-194.

Yelagin, I. N., and Izotov, V. F. (1968). Soil temperatures in the pine forest zone in different seasons. *Sov. Soil Sci.,* 827-831.

Zoltai, S. C., and Tarnocai, C. (1971). Properties of a wooded palsa in northern Manitoba. *Arc. Alp. Res.* **3,** 115-129.

Chapter 2

THE EFFECTS
OF FREEZING

All soils with pergelic temperature regimes and nearly all with cryic temperature regimes are frozen at least part of each year. The only cryic soils that are seldom or never frozen below the upper few centimeters are those of strongly maritime climates with heavy winter precipitation, as in southeastern Alaska and the Aleutian Islands. Freezing involves much more than the conversion of soil moisture to ice. Accompanying this phase transformation are volume changes, migration of moisture, and related phenomena that have lasting effects on soil properties. The degree to which soil properties are altered depends to a large extent on the duration and intensity of freezing and the frequency of freeze–thaw cycles. Many of the consequences of freezing are different in soils with no permafrost or with deep permafrost tables than in soils with ice-rich permafrost at shallow depths. The presence of an essentially impermeable perennially frozen substratum is a special circumstance that results in reactions to freezing and thawing that generally do not occur in soils with permeable substrata.

In soils without permafrost or with permafrost deep enough so that it has little effect on surface processes, freezing begins at the surface at the onset of air temperatures lower than 0° C and except where a thick snow cover completely protects the soil against heat loss, advances downward throughout the winter. During freezing there is a steady migration of soil moisture toward the freezing front and even within the frozen soil itself toward lower temperatures at the surface (Fedorova and Yarilova, 1972; Fedorova, 1974). The amount of moisture moving upward is greatest in poorly drained soils, soils with high water tables, and fine-grained soils with high capillarity (Sokolov and Targul'yan, 1976). The rate and amount of moisture migration also increases as the soil temperature decreases (Pustovoytov, 1962).

I. Moisture Migrations during Freezing

Where soil moisture is generally high, as in western Siberia, the initial rate of advance of the freezing front is less than 1.5 cm per day and by late winter has slowed to between 1 and .5 cm per day. The temperature of the frozen soil fluctuates between 0° and -2° C. At these temperatures thin films of unfrozen water exist in the apparently completely frozen soil (Tsytovich, 1975), and there is migration of moisture from the central to the peripheral parts of soil aggregates and through soil pores toward the colder upper layers. This, along with the heat released during gradual conversion of the film water to ice, keeps temperatures from dropping very far below the freezing point (Fedorova, 1974).

Where soils are not high in moisture, as in areas with dry continental climates, the rate of advance of the freezing front averages 1.4–1.9 cm per day during the winter (Pustovoytov, 1962). Again, it is more rapid in the early part of the cold season. In late winter much of the heat lost to the atmosphere comes from the upper soil layers, which reach very low temperatures. The freezing front gradually approaches a state of equilibrium between heat lost to upper layers and heat generated by the conversion of liquid water to ice. Equilibrium at a temperature of 0° C is eventually attained at the maximum depth of frost penetration and can last as long as 90 days if the freezing front is very deep in dry soils (Fig. 7).

Water that is transported from deep in the soil in response to freezing can be significant in replenishing moisture supplies in the root zone, but appreciable upward water movement occurs only when the underlying soil is moist. Little moisture migration occurs when the soil–water tension is greater than 5 atm (Ferguson *et al.*, 1964).

In the dry climate of southeastern Siberia, 65–70 mm of water migrate into the upper meter of well-drained soils and about 110 mm in poorly drained soils during the winter (Pustovoytov, 1962). In northeastern Siberia the upper part of soils with deep permafrost on floodplains accumulates 40–60 mm of water when the soils are wet at the beginning of the winter, but only 10–20 mm if the soils are dry when they start to freeze (Konorovskiy, 1976). In western North Dakota the water table recedes as the freezing front deepens, and the amount of

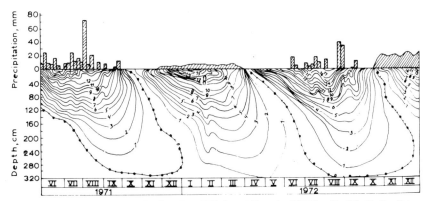

FIGURE 7. *Temperature progression from June 1971 through December 1972 in soils of the Undine Depression, Transbaikal Region, Siberia. Temperatures are in ° C and are negative in the winter months. Precipitation is measured as weekly rainfall in summer and thickness of snow pack in winter. (From O. W. Makeev and A. S. Kerzhentsev [1974]. Cryogenic processes in the soils of northern Asia.* Geoderma 13, *101–109.)*

moisture in the upper meter increases proportionately (Fig. 8). Water accumulated in the root zone in this way makes autumn irrigation of these dry soils unnecessary, (Willis *et al.,* 1964), but much of the moisture is lost to evaporation in the spring (Ferguson *et al.,* 1964).

As noted in the discussion of soil temperatures, the rate of thaw in the spring is more rapid than the freezing rate in the fall. Warming at the surface may begin even before the snow disappears completely because of movement of moisture in vapor form into the soil and release of energy on its subsequent condensation on mineral particles (Longley, 1967). Much of the warming trend, however, is owing to the greater effectiveness of upward movement of moisture from the unfrozen substratum as moisture begins to enter the soil from the surface (Brinkman, 1968). Toward the end of the freezing period, moisture moving upward toward the freezing front supplies enough heat so that the state of equilibrium at 0° C is destroyed, and thawing proceeds from the bottom of the frozen layer as well as from the surface. As a result, the last frost to disappear is in the middle of the frozen zone. In western Siberia soils remain frozen longest at depths of 50–100 cm although the total depth of freezing is about 150 cm (Fedorova, 1964). In dry soils at the forest–steppe transition in central Siberia, the moisture content in the middle part of the soil profile, which remains frozen longest, is consistently higher throughout the summer than either the upper part of the soil or the substratum (Veredchenko, 1959). In very dry soils there is only minimal thawing from the bottom of the frozen zone, and this takes place near the end of the thawing period (Pustovoytov, 1962).

During the spring thaw two saturated zones appear in moist, cold, forested soils, one immediately above the remaining frozen layer and one immediately below it (Kaplyuk, 1964; Fedorova, 1974). The upper zone is simply water

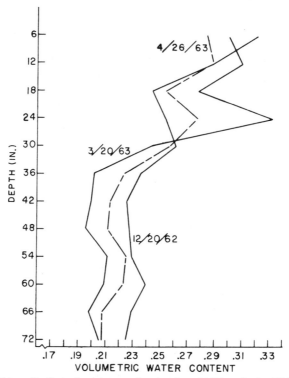

FIGURE 8. *Moisture distribution in a Montana soil on three dates: (a) Prior to freezing (12/20/62); (b) before spring thaw (3/20/63); and (c) after spring thaw (4/26/63). Moisture moves upward to frozen zone in winter and is redistributed downward only slowly in the spring (Ferguson* et al., *1964). (Reprinted from* Soil Science Society of America Proceedings, *Volume 28, page 701, 1964. By permission of the Soil Science Society of America.)*

from melting ice and spring precipitation that is perched above the frozen layer. This water may be well in excess of the field capacity of the soil; in sloping soils that have been cleared of the native vegetation, it is a potential erosion hazard (Izotov, 1968). The lower saturated zone contains water that has migrated from lower layers and melted ice from the bottom of the frozen zone. It may remain in the soil until the profile is completely thawed, which may be well after moisture in the upper saturated zone has been lost to evapotranspiration. All of the excess water drains out of the profile 1 or 2 weeks after complete thawing (Veredchenko, 1959), but the periods of saturation may be long enough so that mottling and other characteristcs of wetness develop in the two zones even in soils with good drainage (Fedorova and Yarilova, 1972).

Soils in low topographic positions may, under certain circumstances, have water under hydrostatic pressure trapped beneath the frozen layer. During the spring thaw this water may be forced into the frozen zone and, in the process, cause distortion and mixing of soil layers (Johnsgard, 1971). Large mounds in poorly drained portions of floodplains are commonly also the result of hydrostatic pressures (Odynsky, 1971).

II. Effects of Deep Freezing

Soils subject to prolonged deep freezing or to frequent freeze–thaw cycles develop characteristics that generally are not found in soils that freeze only to moderate depths. Deep freezing affects soil structure, the distribution of organic matter and clay, and the surface configuration of the soils, and it may also result in internal disruptions. Not all of these occur in every cold soil—some cold soils are, except for temperature, indistinguishable from soils of temperate climates—but most of them exhibit one or more properties that are the direct result of deep or frequent freezing.

A. ICE DISTRIBUTION IN FROZEN SOILS

Ice is seldom uniformly distributed in a frozen soil. It occurs, instead, in patterns that are reflected in soil structure and, in some cases, in the distribution of soil components after the soil has thawed. In the upper part of the soil where plant roots are concentrated and where there are many macropores and root channels, moisture is trapped in place as freezing commences and is frozen in the form of ice blocks or relatively thick veins (Makeev and Kerzhentsev, 1974; Fedorova, 1974). Dissolved gases expelled from the water during freezing are trapped as bubbles in the frozen mass and form smooth-walled vesicular pores that persist in the upper part of the soil after thawing (Harris and Ellis, 1980). The root zone, however, rarely exceeds 50 cm in cold soils and may be as thin as 10 cm. Below this, water migrates into closely spaced undulating lamellae roughly parallel to the soil surface whose appearance in cross section has given rise to such terms as "ice-gneiss" (Taber, 1943) and "sirloin ice" (Everett, 1966). At moderate depths these lenses of ice are generally only about 1 mm thick. At greater depths, where the rate of freezing is not as great, they are thicker and more widely spaced. After thawing, the soil below the root zone retains its thin lamellar or platy structure. The thin zones between the plates are favored sites for the formation of ice lamellae in succeeding years, and, in a relatively short time, the platy structure becomes a prominent feature of the soil below the root zone. This structure is a widely observed phenomenon (Pustovoytov, 1962; Viereck, 1965; Morozova, 1965; Everett, 1966; Yarilova *et al.,* 1970; Bouma and Van der Plas, 1971; Fahey, 1973; Makeev and Kerzhentsev, 1974; Retzer, 1974; Pawluk and Brewer, 1975) and appears to be a

universal characteristic of deeply frozen, loamy soils. Lensing is not common
in gravelly or coarse-sandy soils, in which the ice tends to form films around the
individual grains (Smith, 1956).

B. REDISTRIBUTION OF FINE PARTICLES

In the middle part of the profile of loamy soils, where the soil remains frozen
longest, migration of moisture even at temperatures below the freezing point
results in local redistribution of fine particles. A whitish powdering, which consists principally of quartz and feldspar grains of silt-size, coats the upper surfaces of the platy structural units (Acton and St. Arnaud, 1963; Morozova,
1965). In well-drained soils the color of the upper plate surfaces may be one
value step lighter on the Munsell scale than the interiors of the plates (Rieger *et
al.,* 1963). In some moderately well-drained or somewhat poorly drained soils,
a thin layer of fine clay and iron oxide coats the lower plate surfaces (Fedorova
and Yavilova, 1972). The whitish powdering may be a residue that remains
after water migrating toward the freezing front carries iron in solution and fine
clay in suspension from the bottoms of the thin layers of moisture concentration
between the plates to the upper parts of the layers (Makeev and Kerzhentsev,
1974). The coatings on the lower plate surfaces may result from precipitation of
these materials as freezing proceeds and, eventually, pressure against the lower
faces of the plates by expanding ice. In some finer soils the coatings on both surfaces may simply be precipitates from the soil solution as the concentration of
solutes in the remaining liquid water increases during freezing (Acton and St.
Arnaud, 1963). The whitish powdering becomes less conspicuous with depth,
but roughly horizontal brown bands, presumably iron oxide that has
precipitated after being carried in solution to ice lenses, are common. Horizontal bands of organic matter, believed to be formed by the freezing and fixation
of organic materials in solution or suspension, also occur at various depths in
soils of cold continental areas (Makeev and Kerzhentsev, 1974).

Freezing also affects the distribution of clay in loamy soils by increasing the
concentration of ions in the soil solution. In upper horizons this promotes the
flocculation of clay and the formation of small clay aggregates or floccules.
Flocculation is most pronounced in the presence of highly dispersed humus and
high concentrations of calcium and magnesium ions and in soils that are not
completely saturated when freezing begins (Morozova, 1965). This situation is
most common in the surface horizons of well-drained soils of cold continental
climates. Immediately below the surface horizon but still in the root zone,
where the soil structure is blocky rather than platy, compression and compac-
tion during freezing forces thin films of water containing suspended fine clay
and dissolved iron and organic matter against the coarser soil aggregates and
creates a layer of oriented material on the surfaces of the peds (Gorbunov,

1961; Nogina *et al.*, 1968). This layer resembles the skins, or cutans, formed by deposits from percolating waters that are characteristic of Alfisols (Chapter 6), but the material in this case originates in place rather than in higher soil horizons. Clay in these pressure faces can be distinguished from illuviated clay chiefly by the absence of flow patterns in the cutans.

There is evidence to indicate that fine particles may also migrate before an advancing frost line (Corte, 1962). Where there are zones of slightly higher clay concentration in the parent material, the clay may become concentrated in undulating thin bands. These bands, once formed, may act as filters for percolating clay in the summer and so add further to their clay content (DeMent, 1962).

Concentrations of silty material, in the form of fine granules or intermittent bands, exist in many cold soils, especially those in or adjacent to moraines. They are generally associated with large sand grains or coarse fragments and either coat the coarser particles or fill the interstices between them. The silty concentrations are densely packed and have very low porosity. They are believed to have originated at the time of rapid degradation of permafrost at the close of the Pleistocene era, when voids left by melting ice veins were filled by predominantly silty material from the surrounding matrix, and to have acquired their high bulk density as a result of pressures generated during annual freezing in postglacial times (Collins and O'Dubhain, 1980).

C. COATINGS ON COARSE FRAGMENTS

Stones in the lower horizons of many cold soils have a strongly adhering coating of silt and clay on their upper faces, but the undersides of the stones are either clear or are stained by organic matter of iron and manganese oxides. This contrast between the upper and lower surfaces of stones is seen most often in soils that are moist at the beginning of the freezing season. It is distinctly different from the situation in dry climates, where carbonate or saline crusts on the undersides of stones are common but no coatings exist on the upper faces. In well-drained soils of areas where precipitation is high enough that excess moisture percolates completely through the soil, water tends to accumulate or hang below large fragments. The pressure exerted by expanding ice as the soil freezes "plasters" fine particles in the soil matrix on the upper surfaces of the stones, but this does not occur on the lower surfaces, which are in contact only with water and any substances carried in solution. If the water contains dissolved organic matter or minerals of relatively low solubility, these materials concentrate at the lower stone surface and are irreversibly precipitated on it during the freezing process. If the water contains only highly soluble minerals, nothing remains on the stone after thawing in the spring. In some cases, as in glacial till in maritime areas which is no longer subject to freezing, an upper coating of silt and clay may remain on stones as a relic of a formerly annually freezing environment.

III. Cryoturbation

Many cold soils are subject to physical disturbance as a result of periodic freezing and thawing. The degree of disturbance depends on the soil temperature and moisture regimes and the soil texture. It is greatest in soils underlain by ice-rich permafrost at shallow depth, but severe turbation may also occur in soils that do not have a perennially frozen substratum. Where turbation is severe, it may override the effects of other genetic processes and account in large part for the distribution of organic matter and for other morphological properties of the soil.

The basic cause of soil turbation or heaving is the increase in soil volume on freezing. Very coarse soils that are dry at the beginning of the freezing period are least susceptible to heaving, but soils with as little as 3% by weight of fine particles are subject to volume changes when they freeze. Fine-grained soils with shallow water tables have the greatest heaving potential. The magnitude of the volume increase is commonly greater than would be expected from freezing of the existing soil moisture because of migration of additional water into the freezing zone and the segregation of ice into lenses (Fahey, 1974). The vertical expansion of a clayey soil with an abundant supply of moisture can be as much as 60% (Viereck, 1965). The intensity of frost heaving is greater where there is a textural interface, such as clay above sand or gravel, within or just below the freezing zone because this tends to increase the moisture supply (Troll, 1958). Heaving also tends to be greater where the vegetative cover is sparse and there is little or no snow cover, largely because soil temperatures are lower under those conditions.

A. FREEZE–THAW CYCLES

Soils subject to repeated cycles of freezing and thawing tend to develop an irregular surface microrelief. The degree of irregularity is related to the intensity of heaving and the frequency of freeze–thaw cycles. It is least in well-drained soils of areas with relatively mild climates that do not freeze deeply and in soils of moderately cold climates that, once frozen in early winter, do not thaw until the following spring. In its extreme form the soil surface consists of closely spaced mounds or hummocks separated by narrow troughs (Fig. 9). The mode of origin and the shape of the mounds are different in the different climatic regions.

Under high mountain climates, especially at lower latitudes, soils have

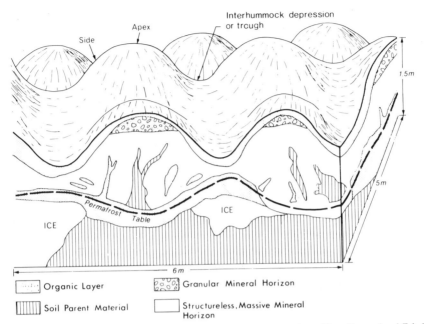

FIGURE 9. *Morphology of earth hummocks in an arctic soil with permafrost. (From Tarnocai and Zoltai [1978] by permission of University of Colorado and the Canadian government.)*

many freeze–thaw cycles in the course of a year. In mountains of East Africa, close to the equator, there is almost daily freezing and thawing at elevations higher than 4000 m (Frei, 1978). In the Snowy Mountains of southeastern Australia (lat 36° S), there are up to 200 freeze–thaw cycles annually at an elevation of about 1900 m (Costin and Wimbush, 1973). Air temperature measurements in the Front Range of Colorado (lat 40° N) suggests 238 freeze–thaw cycles over a 22-month period at 2600 m, but only 89 at 3750 m (Fahey, 1973). The number of cycles decreases at higher elevations and toward the poles because of the increasing severity and length of the winters during which no diurnal thawing and refreezing take place. There are only 9 cycles annually in the arctic islands of Canada—lat 80° N—(Fraser, 1959). In strongly maritime climates, such as those of Iceland and the Aleutian Islands, freezing and thawing cycles occur mainly during the winter months, but the number of cycles may be large.

B. EFFECTS ON SOIL TEXTURE

The grinding action in soils with frequent freeze–thaw cycles may increase the proportion of particles of fine clay size in the soil. These fine particles retain many of the characteristics of their parent materials, but chemical weathering may be accelerated to the point that the particles take on some of the attributes of amorphous clays (Milestone and Wilson, 1977). They are similar to, but not identical with, amorphous clays that occur commonly in Andepts (Chapter 8) and Spodosols (Chapter 5).

Frost action and the grinding movements it generates in soils foster the mechanical disintegration of sand and coarse fragments. Grains produced by this action are dominantly in the size range of .1–.01 mm (Troll, 1958)—that is, in the silt and very fine sand particle-size classes. Because of low soil temperatures, chemical weathering that results in clay-size particles is less effective in cold than in temperate soils although some does occur, especially in mountain soils with frequent freeze–thaw cycles. In general, clayey soils in cold regions form only from clayey parent materials.

Soils of very cold areas commonly have a surface mantle of frost-rived gravel. Even small amounts of moisture in fine fissures in stones can exert tremendous pressures, as much as 2000 kg/cm^2 at -22° C, when frozen (Fraser, 1959). At very low temperatures the stones shatter to produce angular grains with sharp edges. In areas where there is little vegetation or snow cover to protect the soil, as in the polar desert, winds remove any fine material from the upper few centimeters, leaving a pediment that consists almost entirely of shattered coarse fragments (Tedrow, 1966).

C. MOUNDING AND FISSURING

Because few soils are perfectly homogenous in texture, frequent freezing and thawing results in irregular heaving and, commonly, in the formation of a network of soil cracks as much as 60 cm deep. During thaws water flows into the cracks and, on refreezing, expands and widens them (Costin and Wimbush, 1973). Pressure exerted by the ice in the cracks forces the intervening soil upward and, in the process, may move stones and other substratum materials toward the surface. Mounds formed in this way tend to have a conical shape. As the mounds increase in size, the heaving action and lateral soil movements may disrupt the vegetative cover and create bare spots on the surface. These may subside during the thaw portion of the cycle to form a soupy mud spot or "flag" or may remain as a barren area at the top of the mound (Johannesson, 1960).

During freezing periods surface temperatures beneath the bare spots are lower than those under the remaining vegetation. Moisture moves laterally into the soil under the spot in response to the temperature differential, then upward to the surface where it freezes in the form of thin vertical crystals up to 10 cm in length called "needle ice" (Troll, 1958). The crystals, as they form, can tear roots and lift stones to the surface, thus retarding revegetation. Needle ice also forms at the sides of the mounds and, by erosion of the loosened materials during thaws, helps to widen and deepen the existing troughs (Costin and Wimbush, 1973). The crystals may be attached at their base to thin ice lenses in the underlying soil (Fahey, 1973), but ice lenses form only after the needle ice has ceased to grow (Everett, 1966). A vesicular surface structure caused by needle ice formation is common in the bare spots (Troll, 1958).

The extent of mounding, even under frequent freeze–thaw conditions, is determined largely by soil texture and moisture content. Very coarse and excessively drained soils, in which there is little capillary movement of moisture, are seldom mounded, but mounding is common in silty and clayey soils. In Iceland, where volcanic ash is the universal parent material and nearly all virgin soils have surface mounds or "thufur," the height of the mounds is correlated with the amount of moisture available. Mounds are 30–40 cm high in well-drained soils, 35–45 cm high in moderately well-drained soils, and 40–50 cm high in somewhat poorly drained soils. Very poorly drained soils that are under water for long periods, however, do not become mounded. Mounds occur in organic soils that are not flooded, but the height of the mounds in peats is not as great as in mineral soils (Johannesson, 1960).

In forested soils of continental climates with very cold winters and low winter precipitation, deep frost fissures are common. As frozen soils cool, films of liquid water that are present even at subfreezing temperatures and that provide some degree of elasticity disappear; the soils become very rigid and eventually crack as a result of internal stress. A system of fissures develops and forms a network of straight-sided polygons. Each crack, once formed, is filled in the spring with water carrying bits of litter and other mineral and organic material in solution or suspension, and so remains a zone of weakness and a preferential site for fissuring in succeeding years (Fig. 10). The fissures may reach depths of 2 m or more; if there is permafrost within that depth, they may extend below the permafrost table (Gugalinskaya and Alifanov, 1979). Erosion of the upper part of the fissures by water flowing into them or through the channels they create on the surface results in a microrelief of relatively flat-topped mounds 10–50 cm higher than the intervening troughs. This type of mound development is seen chiefly in moderately sloping, well-drained soils (Dimo, 1965; Makeev and Kerzhentsev, 1974).

The inflow of surface materials may result in deep tongues of humus bordering the fissures (Yelovskaya, 1965; Makeev and Kerzhentsev, 1974; Sokolov and Tursina, 1979) (Fig. 11). Some of this organic accumulation may also be the result of migration of dissolved humus toward the open cracks during the winter. The development of deep tongues of humus is more pronounced in soils with permafrost and occurs only intermittently in the warmer parts of the boreal belt. A network of finer vertical cracks in which roots may penetrate to fairly great depth may also develop between the major fissures. This can produce a prismatic structure with thin organic linings on walls of the prisms,

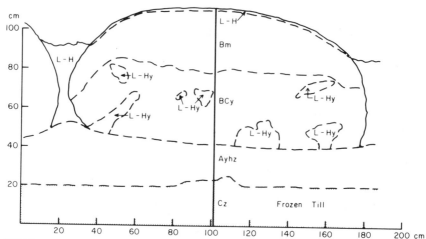

FIGURE 10. *Cross section of a hummock in well-drained, clayey soil of the Mackenzie River area, Northwest Territories, Canada. The symbol L–H indicates organic litter; the Bm horizon has a grayish brown color; the BC horizon has a dark gray color; and the Ayhz horizon contains mixed organic and mineral material. Subscript y indicates cryoturbation, and subscript z indicates a frozen condition.* (From C. Tarnocai [1973]. *Soils of the Mackenzie River Area.* Env.-Soc. Comm., Northern Pipelines, Task Force on Northern Oil Devel., Rept. 73-26, p. 77.)

similar in outward appearance to soils formed under highly saline conditions (Sokolov and Tursina, 1979).

IV. Effects of Shallow, Ice-rich Permafrost on Turbation

In soils with ice-rich permafrost, freezing in late summer progresses not only from the surface downward, but also upward from the permafrost table (Yelovskaya, 1965; Kallio and Rieger, 1969; Kelley and Weaver, 1969; Mackay and MacKay, 1974). At the end of the summer, liquid water in contact with the permafrost no longer transmits heat from the surface but instead loses heat to the permafrost and freezes. The temperature of the frozen soil below the permafrost table remains constant until the overlying soil is completely frozen. After that, it drops rapidly (Mackay and MacKay, 1974). The advance of the freezing front is generally more rapid from the surface, which at this time is colder than the permafrost, than upward from the permafrost table (Yelovskaya, 1965). In well-drained arctic soils with shallow permafrost, however, the response at depth to surface temperature variations may be quite rapid, and substantial upward freezing from the permafrost table takes place almost immediately after a drop in air temperature (Tarnocai, 1980).

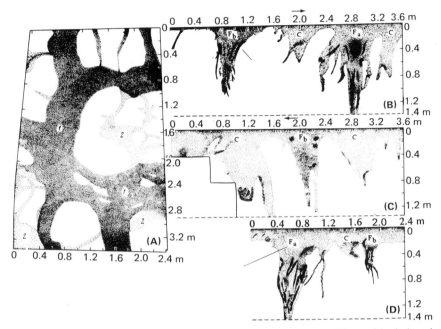

FIGURE 11. *Humus tongues in a soil of the Nercha Basin, in southeastern Siberia. Diagram A is a horizontal section at a depth of 55 cm, and diagrams B, C, and D are vertical sections. Tongues labeled "C" ("1" in diagram A) consist of older, more compact humus and were formed well above the permafrost table. Tongues labeled "F_a" and "F_b" ("2" in diagram A) are more friable and formed in cracks extending to the existing or an earlier permafrost table. The present permafrost table is indicated by broken lines. (From L. A. Gugalinskaya and V. M. Alifonov [1979]. Morphogenetic profile analysis as a basis for the reconstruction of soil formation conditions [as exemplified by the permafrost soils of the Nercha Basin]. Sov. Soil Sci. 11, 261–273. By permission of Scripta Publishing Co.)*

A. HYDROSTATIC MOUNDING

During freezing of the active layer (the soil above the permafrost that thaws in the summer), moisture migrates to both freezing fronts, dehydrating in part the intermediate unfrozen zone (Kaplyuk, 1964). At Barrow, Alaska, a soil that was saturated throughout the summer with about 25% moisture had a moisture content of only 15% between the freezing fronts in early winter (Brown, 1967). In very wet soils and in other soils in years with exceptionally high late-summer rainfall, enough moisture may remain in the soil to be trapped in pockets as the two fronts meet (Tarnocai and Zoltai, 1978). The hydrostatic pressure in these pockets may be great enough to raise the soil above them, or there may be enough additional moisture supplied to ice lenses forming around them so that expansion pressures create a mound at the surface. Low mounds and hummocky surface microrelief probably created in this way are common in tundra areas with maritime climates and are larger and even more prevalent in northern forests where the permafrost table is somewhat deeper (Tarnocai and Zoltai, 1978). In low-lying, very wet tundra areas, hydrostatic pressures result in the development of large mounds, or pingos, with a core made up almost exclusively of ice (Mackay, 1962). Other pingos, in valleys and basins, are formed by the upwelling and subsequent freezing of water under hydraulic pressure in unfrozen seams, or taliks, within the permafrost (Price, 1972).

B. TURBATION ACCOMPANYING ICE WEDGE FORMATION

Soils with shallow permafrost that are saturated or close to the saturation point throughout the summer are by far the most extensive soils in areas north of the arctic tree line (Rieger, 1974). They exist also in alpine tundra areas, where they are moderately extensive (Retzer, 1974), and in association with well-drained forested soils in much of the boreal climatic belt. An important cause of turbation in these soils, especially on level to moderately sloping land in the arctic, is the growth of ice wedges in the underlying permafrost. Frost fissures that reach well into the permafrost develop very early in the soil's history and remain open during the early summer (Kerfoot, 1972) (Fig. 12). Meltwater containing finely divided organic matter and other material flows into the fissures much as it does in fissured forested soils with deep or no permafrost. When this water refreezes and expands, the frozen soil between the fissures is forced upward. As this process, together with increasing erosion of the soil adjacent to the fissures, is repeated year after year wedgelike masses of ice are formed within the permafrost, and the characteristic microrelief of upland tundra areas, consisting of elevated polygons separated by depressions or troughs above the ice wedges, is developed (Fig. 13). Polygons in the arctic tundra are generally much larger than those formed by fissuring in forested areas and may reach diameters of 30 m or more. As the ice wedges grow, the troughs above them widen and, in very old polygonal areas, may attain widths of several meters (Bird, 1974). The surface microrelief of the polygons is generally irregular and, especially in clayey soils, may include frost circles in which the vegetative cover has been disrupted.

Some low-lying, very wet soils, such as those developing in recently deposited sediments of deltas and other coastal areas, have polygons in which the rims are higher than the polygon centers because of upheaval over newly formed ice wedges. In time, as the land is elevated, these low-centered polygons are converted to flat-centered and high-centered polygons (Drew, 1957).

Accompanying the growth of ice-wedge polygons is a slow, intermittent movement of soil material that resembles a churning or convective process. Organic material from the upper part of the soil that is carried along in this movement accumulates in tongues beneath the troughs between polygons

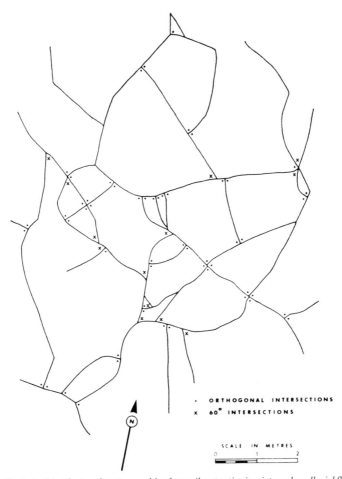

ORTHOGONAL INTERSECTIONS

x   60° INTERSECTIONS

SCALE   IN   METRES

0   1   2

FIGURE 12. *An incipient frost crack pattern resulting from soil contraction in winter on low alluvial flats in the Mackenzie River delta. Primary frost cracks generally are oriented at right angles to a body of water; secondary cracks tend to propagate toward the primary cracks and to intersect them at an angle of 90°. Intersection angles of 60° occur mostly as a result of bifurcation of a primary frost crack. This incipient pattern is preserved in the later development of ice-wedge polygons. (From Kerfoot [1972] by permission of the Arctic Institute of North America.)*

(Mackay *et al.*, 1961; Ignatenko, 1963; James, 1970; Pettapiece, 1975). Some of the organic matter diffuses from the tongues into pockets and streaks under the polygon centers (Leahey, 1947; Pettapiece, 1975), but much of it accumulates and spreads laterally at the permafrost table, which apparently acts as a floor to the churning process. This accumulation of fibrous or partially decomposed organic matter in a discontinuous layer at the permafrost table has been observed widely in soils with shallow permafrost under both tundra and forest vegetation (Karavayeva and Targul'yan, 1960; Mackay *et al.*, 1961;

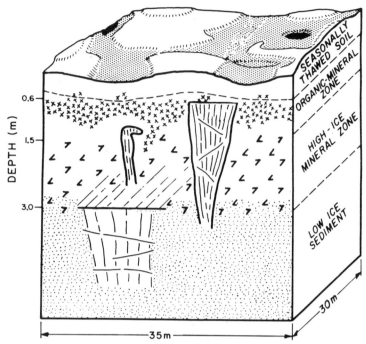

FIGURE 13. *Diagram indicating the relationship between ice wedges and polygons in tundra soils. The upper surface of the truncated ice wedge represents the depth of thaw during an earlier warm interval. Secondary ice wedges formed during refreezing of sediment to the present level of permafrost table (Brown, 1967). (Reprinted from* Soil Science Society of America Proceedings, *Volume 31, page 688, 1967. By permission of the Soil Science Society of America.)*

Ignatenko, 1963; Day and Rice, 1964; Ivanova, 1965; Ignatenko, 1966; Archegova, 1974; Pettapiece, 1974; Zoltai and Pettapiece, 1974) and can be considered to be a typical feature of these soils. In many cases there is organic accumulation both above and below the permafrost table; the deeper material is believed to have been deposited during an earlier, warmer period and to have been trapped below the present, shallower permafrost table (Brown, 1965). It is likely that much of the organic matter, once transported to the permafrost table, does not return to the surface because it is below the zone of maximum frost stirring (Mackay *et al.*, 1961).

V. Intensity of Cryoturbation

Factors that influence the intensity of frost stirring are soil texture, soil wetness, the thickness of the organic mat at the surface, and the depth to the permafrost table, if any. In soils with permafrost, churning is more active in fine-grained than in coarse material, and related phenomena, such as the layer of organic matter at the permafrost table, are therefore better developed in the finer soils. In general there is greater mechanical disturbance and greater organic accumulation at the permafrost table in soils that have moisture contents near the saturation point throughout the summer than in soils with lower moisture contents (Ignatenko, 1963), but soils with standing water on the surface may be less disturbed. Where the permafrost table is deeper than about 1 m, as in the maritime tundras of eastern Europe, organic material may not be carried deeply enough to form a layer above the permafrost (Karavayeva and Targul'yan, 1960). If the organic mat on the soil surface is very thick, as in many soils with tundra vegetation in the boreal climatic belt, the permafrost table may be at or near the base of the organic mat. Frost stirring in such soils is less significant, and a polygonal surface relief may not have developed. Humus accumulation at deep permafrost tables in well-drained, forested soils has been observed, but this probably is more the result of migration of dissolved humus than of stirring processes (Dimo, 1965).

A. HUMMOCKS

In forested soils of the coldest parts of the boreal belt, upwelling results in hummocks as much as 1 m high and 4 m wide (Viereck, 1965). Streaks and smears of organic matter and distorted horizons in the hummocks, as well as tipped trees on their sides, are evidence of intensive soil disturbance. It is likely that activity in these soils is intermittent; soils that have been in a stable state long enough for genetic horizons to develop may be disrupted after a series of wet years or during the years following a severe forest fire when the permafrost table is initially lowered and then rises as vegetation is reestablished (Zoltai and Pettapiece, 1974). A major period of hummock formation appears to have occurred about 4500 years ago during a period of climatic change toward cooler and more humid conditions. Many hummocks formed at that time have now become relatively stable (Zoltai *et al.*, 1978).

B. FROST CIRCLES

Upward movement of the soil mass in the centers of polygons can result in nearly perpendicular orientation of the long axes of rock fragments (Kreida, 1958) and of originally horizontal strata, and in disruption of the vegetative cover (Vasil'yevskaya, 1979). The proportion of surface area occupied by barren frost circles is higher in colder areas, partly because of the thinner plant cover in those areas and partly because of more intense frost heaving. Stones tend to accumulate at the surface of frost circles because of heaving and the removal of fine material from the bare surface by winds. Some of the fine material is carried by runoff water into the troughs between polygons (Ignatenko, 1963). The growth of needle ice in the early winter makes for a slightly convex surface in the frost circles; stones lifted by the ice crystals fall toward the edges of the circles and eventually form a garland around them (Troll, 1958; Everett, 1966).

VI. Effects of Freezing on Sloping Soils

On slopes generally between 4° and 12° (7–21%) but in places on gradients as low as 2° (3.5%), saturated, fine-grained soils move downslope in a complex and irregular manner that usually results in a pattern of lobate sloping terraces. This soil flow, or solifluction, may occur in any cold soil but is most prominent under tundra vegetation in soils with shallow permafrost tables. Most of the flow takes place in the spring before moisture from thawing ice has run off (Smith, 1956; Bird, 1974). Because maximum movement is in the upper part of the soil, there is a continual overturning of soil material. This results in buried organic layers, irregular textural boundaries, tongues of intruding materials, and an accumulation of stones near the front of the lobes (Ugolini, 1966; Harris and Ellis, 1980). Slips commonly create cracks and semicircular bare patches in the back portions of the lobes (Bird, 1974). Once formed, these patches are maintained by intense heaving and needle ice formation (Smith, 1956).

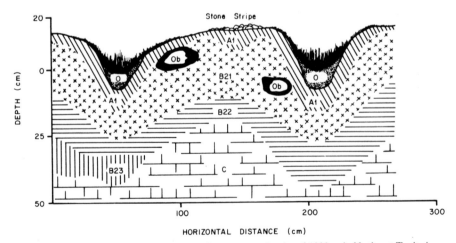

FIGURE 14. *Cross section through a stone stripe area at an elevation of 1000 m in Northwest Territories, Canada. Incorporated organic material at the edges of the stripe is near the bottom of the zone of maximum cryoturbation (Walmsley and Lavkulich, 1975).* (Reprinted from Soil Science Society of America Proceedings, *Volume 29, pages 84–88, 1975. By permission of the Soil Science Society of America.)

In tundra areas, all soils of steep slopes, well-drained soils of lower slopes, and ridge tops that have a sparse vegetative cover are highly susceptible to denudation by frost action. These soils commonly have a high proportion of unvegetated areas. Textural sorting, with rings of coarse fragments around frost circles, is common on the surface of level to moderately sloping soils. On steep slopes the coarse fragments become aligned in a series of stripes that run vertically down the slope and terminate in concentric lobes of gravel as the gradient decreases to about 15° (Fig. 14). Fine-grained soils with few stones may not be texturally sorted but may exhibit a series of vegetated stripes separated by nearly bare ground. These are believed to originate from single plants established high on the slope that create a stable environment below them, thus permitting down-slope colonization. Soil particles moving downslope on either side prevent lateral extension of the vegetation (Smith, 1956).

References

Acton, D. F., and St. Arnaud, R. J. (1963). Micropedology of the major profile types of the Weyburn catena. *Can. J. Soil Sci.* **43**, 377–386.

Archegova, I. B. (1974). Humus profiles of some taiga and tundra soils in the European U. S. S. R. *Sov. Soil Sci.* **6**, 136–141.

Bird, J. B. (1974). Geomorphic processes in the Arctic. *In* "Arctic and Alpine Environments" (J. D. Ives and R. G. Barry, eds.), pp. 703–720. Methuen, London.

Bouma, J., and Van der Plas, L. (1971). Genesis and morphology of some alpine pseudogley profiles. *J. Soil Sci.* **22**, 81–93.

Brinkman, W. A. R. (1968). Some comments on the paper by R. W. Longley. *Soil Sci.* **106** 469–470.

Brown, J. (1965). Radiocarbon dating, Barrow, Alaska. *Arctic* **18**, 37–48.

Brown, J. (1967). Tundra soils formed over ice wedges, northern Alaska. *Soil Sci. Soc. Am. Proc.* **31**, 686–691.

Collins, J. F., and O'Dubhain, T. (1980). A micromorphological study of silt concentrations in some Irish Podzols. *Geoderma* **24**, 215–224.

Corte, A. E. (1962). Vertical migration of particles in front of a moving freezing plane. *J. Geophys. Res.* **67**, 1085–1089.

Costin, A. B., and Wimbush, D. J. (1973). Frost cracks and earth hummocks at Kosciusko, Snowy Mountains, Australia. *Arc. Alp. Res.* **5**, 111–120.

Day, J. H., and Rice, H. M. (1964). The characteristics of some permafrost soils in the Mackenzie Valley, N. W. T. *Arctic* **17**, 223–236.

DeMent, J. A. (1962). "The Morphology and Genesis of the Subarctic Brown Forest Soils of Central Alaska." Unpub. Ph.D. Dissertation, Cornell U.

Dimo, V. N. (1965). Formation of a humic-illuvial horizon in soils in permafrost. *Sov. Soil Sci.*, 1013–1021.

Drew, J. V. (1957). "A Pedologic Study of Arctic Coastal Plain Soils Near Point Barrow, Alaska." Unpub. Ph.D. Dissertation, Rutgers U.

Everett, K. R. (1966). Slope movement and related phenomena. *In* "Environment of the Cape Thompson Region, Alaska" (N. J. Wilimovsky, ed.), pp. 175–220. U. S. Atomic Energy Comm., Div. Tech. Inf. Ext., Oak Ridge, Tennessee.

Fahey, B. D. (1973). An analysis of diurnal freeze-thaw and frost heave cycles in the Indian Peaks region of the Colorado Front Range. *Arc. Alp. Res.* **5**, 269-281.

Fahey, B. D. (1974). Seasonal frost heave and frost penetration measurements in the Indian Peaks region of the Colorado Front Range. *Arc. Alp. Res.* **6**, 63-70.

Fedorova, N. M. (1974). Thermal and moisture regimes in a soil profile affected by prolonged seasonal freezing (middle taiga subzone, West Siberia). *Geoderma* **12**, 111-119.

Fedorova, N. M., and Yarilova, E. A. (1972). Morphology and genesis of prolonged seasonally frozen soils of western Siberia. *Geoderma* **7**, 1-13.

Ferguson, H., Brown, P. L., and Dickey, D. D. (1964). Water movement and loss under frozen soil conditions. *Soil Sci. Soc. Am. Proc.* **28**, 700-703.

Fraser, J. K. (1959). Freeze-thaw frequencies and mechanical weathering in Canada. *Arctic* **12**, 40-53.

Frei, E. (1978). Andepts in some high mountains of East Africa. *Geoderma* **21**, 119-131.

Gorbunov, N. I. (1961). Movement of colloidal and clay particles in soils. *Sov. Soil Sci.,* 712-724.

Gugalinskaya, L. A., and Alifanov, V. M. (1979). Morphogenetic profile analysis as a basis for the reconstruction of soil formation conditions (as exemplified by the permafrost soils of the Nercha Basin). *Sov. Soil Sci.* **11**, 261-273.

Harris, C., and Ellis, S. (1980). Micromorphology of soils in soliflucted materials, Okstindam, northern Norway. *Geoderma* **23**, 11-29.

Ignatenko, I. V. (1963). Arctic tundra soils of the Yugor Peninsula. *Sov. Soil Sci.,* 429-440.

Ignatenko, I. V. (1966). Vaygach Island soils. *Sov. Soil Sci.,* 991-998.

Ignatenko, I. V. (1971). Soils in the Kara River basin and their zonal position. *Sov. Soil Sci.* **3**, 15-28.

Ivanova, Ye. N. (1965). Frozen taiga soils of northern Yakutia. *Sov. Soil Sci.,* 733-744.

Izotov, V. F. (1968). Pattern of soil freezing and thawing in the waterlogged forests of the northern taiga subzone. *Sov. Soil Sci.,* 807-813.

James, P. A. (1970). The soils of the Rankin Inlet area, Keewatin, N. W. T., Canada. *Arc. Alp. Res.* **2**, 293-302.

Johannesson, B. (1960). "The Soils of Iceland." Univ. Res. Inst., Dept. Agr. Rep. B-13, Reykjavik.

Johnsgard, G. A. (1971). Pedoturbation by artesian action. *Soil Sci. Soc. Am. Proc.* **35**, 612-616.

Kallio, A., and Rieger, S. (1969). Recession of permafrost in a cultivated soil of interior Alaska. *Soil Sci. Soc. Am. Proc.* **33**, 430-432.

Kaplyuk, L. F. (1964). Some properties of gleyey-slightly podzolic soils in the northern Ob' region in the Salakhard Rayon. *Sov. Soil Sci.,* 36-43.

Karavayeva, N. A., and Targul'yan, V. O. (1960). Humus distribution in the tundra soils of northern Yakutia. *Sov. Soil Sci.,* 1293-1300.

Kelley, J. J., and Weaver, D. F. (1969). Physical processes at the surface of the arctic tundra. *Arctic* **22**, 425-437.

Kerfoot, D. E. (1972). Thermal contraction cracks in an arctic tundra environment. *Arctic* **25**, 142-150.

Konorovskiy, A. K. (1976). Characteristics of formation and properties of the floodplain soils of central Yakutiya. *Sov. Soil Sci.* **8**, 125-132.

Kreida, N. A. (1958). Soils of eastern European tundras. *Sov. Soil Sci.,* 51-56.

Leahey, A. (1947). Characteristics of soils adjacent to the Mackenzie River in the Northwest Territories of Canada. *Soil Sci. Soc. Am. Proc.* **12**, 458-461.

Longley, R. W. (1967). Temperature increases under snow during a mild spell. *Soil Sci.* **104**, 379-382.

Mackay, J. R. (1962). Pingos of the Pleistocene Mackenzie Delta area. *Can. Dept. Mines and Tech. Surv., Geog. Bull. 18,* 21-63.

Mackay, J. R., and MacKay, D. K. (1974). Snow cover and ground temperatures, Garry Island, N. W. T. *Arctic* **27**, 287–296.

Mackay, J. R., Mathews, W. H., and MacNeish, R. S. (1961). Geology of the Engigstciak archaeological site, Yukon Territory. *Arctic* **14**, 25–52.

Makeev, O. W., and Kerzhentsev, A. S. (1974). Cryogenic processes in the soils of northern Asia. *Geoderma* **12**, 101–109.

Milestone, N. B., and Wilson, A. T. (1977). Amorphous constituents of some high altitude soils of the South Island of New Zealand. *J. Soil Sci.* **28**, 379–391.

Morozova, T. D. (1965). Micromorphological characteristics of Pale Yellow permafrost soils of central Yakutia in relation to cryogenesis. *Sov. Soil Sci.*, 1333–1342.

Nogina, N. A., Lebedeva, I. I., and Shurygina, Ye. A. (1968). Effect of subzero temperatures on the solubility and mobility of nonsilicate iron. *Sov. Soil Sci.*, 1660–1667.

Odynsky, W. (1971). Mounds (humpies) in the Peace River area of Alberta. *Can. J. Soil Sci.* **51**, 132–135.

Pawluk, S., and Brewer, R. (1975). Micromorphological and analytical characteristics of some soils from Devon and King Christian Islands, N. W. T. *Can. J. Soil Sci.* **55**, 349–361.

Pettapiece, W. W. (1974). A hummocky permafrost soil from the subarctic of northwestern Canada and some influences of fire. *Can. J. Soil Sci.* **54**, 343–355.

Pettapiece, W. W. (1975). Soils of the subarctic in the lower Mackenzie basin. *Arctic* **28**, 35–53.

Price, D. W. (1972). "The Periglacial Environment, Permafrost, and Man." Assn. Am. Geog. Res. Pap. 14.

Pustovoytov, N. D. (1962). Influence of seasonal freezing on the water regime of Amur soils. *Sov. Soil Sci.*, 575–583.

Retner, J. L. (1974). Alpine soils. *In* "Arctic and Alpine Environments" (J. D. Ives and R. G. Barry, eds.), pp. 771–802. Methuen, London.

Rieger, S. (1974). Arctic soils. *In* "Arctic and Alpine Environments" (J. D. Ives and R. G. Barry, eds.), pp. 749–769. Methuen, London.

Rieger, S., DeMent, J. A., and Sanders, D. (1963). "Soil Survey of Fairbanks Area, Alaska." U. S. Dept. Agr., Soil Cons. Serv., Ser. 1959, No. 25. Washington, D.C.

Smith, J. (1956). Some moving soils in Spitsbergen. *J. Soil Sci.* **7**, 10–21.

Sokolov, I. A., and Targul'yan, V. O. (1976). Taiga soils of Transbaykalia in relation to the originalness of soils in the permafrost taiga region. *Sov. Soil Sci.* **8**, 405–416.

Sokolov, I. A., and Tursina, T. V. (1979). Pale Yellow-Gray soils of central Yakutia, an analog of Gray Forest soils. *Sov. Soil Sci.* **11**, 125–136.

Taber, S. (1943). Perennially frozen ground in Alaska: Its origin and history. *Geol. Soc. Am. Bull.* **54**, 1433–1548.

Tarnocai, C. (1973). "Soils of the Mackenzie River Area." Envir.-Social Comm., Northern Pipelines, Task Force on Northern Oil Devel. Rep. 73-26. Ottawa, Ontario.

Tarnocai, C. (1980). Summer temperatures of Cryosolic soils in the north-central Keewatin, N. W. T. *Can. J. Soil. Sci.* **60**, 311–327.

Tarnocai, C., and Zoltai, S. C. (1978). Earth hummocks of the Canadian arctic and subarctic. *Arc. Alp. Res.* **10**, 581–594.

Tedrow, J. C. F. (1966). Polar desert soils. *Soil Sci. Soc. Am. Proc.* **30**, 381–387.

Troll, C. (1958). "Structure Soils, Solifluction, and Frost Climates of the Earth." U. S. Army Snow, Ice, and Permafrost Res. Estab. (SIPRE) Trans. **43**. Hanover, New Hampshire.

Tsytovich, N. A. (1975). "The Mechanics of Frozen Ground." Scripta, Washington, D.C.

Ugolini, F. C. (1966). Soils of the Mesters Vig District, northeast Greenland: II. Exclusive of Arctic Brown and Podzol-like soils. *Medd. om Gron.*, Bd. **176**, *No. 2*, 25 pp.

Vasil'yevskaya, V. D. (1979). Genetic characteristics of soils in a spotty tundra. *Sov. Soil Sci.* **11**, 390–401.

Veredchenko, Yu. P. (1959). Characteristics of the physical properties, water, and temperature of Krasnoyarsk forest-steppe soils. *Sov. Soil Sci.,* 1031–1040.

Viereck, L. A. (1965). Relationship of white spruce to lenses of perennially frozen ground, Mount McKinley National Park, Alaska. *Arctic* **18,** 262–267.

Walmsley, M. E., and Lavkulich, L. M. (1975). Landform-soil-vegetation-water chemistry relationships, Wrigley area, N. W. T.: I. Morphology, classification, and site description. *Soil Sci. Soc. Am. Proc.* **39,** 84–88.

Willis, W. O., Parkinson, H. L., Carlson, C. W., and Haas, H. J. (1964). Water table changes and soil moisture loss under frozen conditions. *Soil Sci.* **98,** 244–248.

Yarilova, Ye. A., Makeyev, O. V., and Tsybzhitov, Ts. Kh. (1970). Genetic and micromorphological characteristics of some soils in the western Baikal region. *Sov. Soil Sci.* **2,** 207–219.

Yelovskaya, L. G. (1965). Saline soils of Yakutia. *Sov. Soil Sci.,* 355–359.

Zoltai, S. C., and Pettapiece, W. W. (1974). Tree distribution on perennially frozen earth hummocks. *Arc. Alp. Res.* **6,** 403–411.

Zoltai, S. C., Tarnocai, C., and Pettapiece, W. W. (1978). Age of cryoturbated organic materials in earth hummocks from the Canadian Arctic. *Proc. 3rd Int. Conf. on Permafrost, Vol* **1,** 326–331.

Chapter 3

THE UNITED
STATES SOIL
TAXONOMY

Increasing dissatisfaction with the system of soil classification that had been in use since 1938 (Baldwin *et al.,* 1938) led leaders of the soil survey staff in the United States to begin work in 1951 on a new and more precise classification system that would provide better definitions and groupings of soils and make possible more reliable interpretations for use (Kellogg, 1963). Some basic principles used in development of the new system were:

1. Soils are to be defined in terms of their own properties rather than by external factors such as climate, geology, and the vegetation they support, although it is recognized that these factors have a profound influence on soil characteristics. (It was recognized also that the temperature and moisture regimes of soils are important soil properties, related to, but distinct from, climatic characteristics.)
2. Soil characteristics related to their genesis are to be used to the fullest extent possible to make distinctions between soils, and preference is to be given to those characteristics that can be observed in the field.
3. The number of genetic characteristics used as distinguishing criteria, including those inherited from the parent materials, is to increase at successively lower categories so that at the lowest level all of the genetic features of a soil are recognized.

4. All classes are to be defined quantitatively in terms of measurable soil properties rather than in genetic terms, thus ensuring that any given soil can be classified in only one way (Smith, 1963).

Development of the new classification scheme proceeded by means of a series of approximations, each of which was tested in the field and revised as appropriate to better realize the objectives of the system. The seventh approximation, particularly, was extensively tested and was adopted in 1965 for use in the United States soil survey. After still more refinement the new scheme was published as the official United States Taxonomy in 1975.

I. Pedons and Polypedons

Among the concepts introduced during the development of the United States Taxonomy is that of the *pedon,* which is defined as the smallest area suitable for description and sampling as a means of characterizing larger bodies of soil. A pedon has no fixed boundaries, but it must be large enough to adequately represent any variability in the properties of horizons within the larger body, yet small enough to permit detailed examination of the soil. Where horizons are continuous and uniform, a surface area as small as 1 m^2 can satisfy these requirements. Where preliminary examination of the larger area indicates that horizons are intermittent or vary in thickness, at least a portion of the recurrent variability must be covered, but the pedon or sampling unit should not have a surface area larger than 10 m^2. In such soils several pedons may need to be examined to determine the full range of variability within a given soil unit.

The soil unit, or individual, that is actually classified is an assemblage of adjoining pedons called a *polypedon.* Delineations on detailed soil maps are, ideally, boundaries between polypedons of different kinds. In practice these boundaries more often than not represent gradations between different kinds of soils rather than abrupt transitions, and delineated areas generally include pedons or small polypedons of other soils. The name applied to the delineated area, however, is that of the polypedon or, in the case of complex associations, polypedons that are dominant within it. Field investigators of soils have always been aware, consciously or otherwise, of the relationships between sampling units and the soil boundaries that are shown on maps, but the concept of pedons and polypedons provides the theoretical basis for the link between the classification system and its practical application in the field (Johnson, 1963).

II. Diagnostic Horizons

Another new concept is that of diagnostic horizons, or readily recognizable soil layers that are the products of genetic processes. Because the diagnostic horizons are the result of processes that have actually taken place, their identification makes possible classification based on soil genesis without reliance on assumptions tied to climatic regimes or the native vegetation. They are divided into two groups—surface horizons, or epipedons, and subsurface horizons. The horizons are not necessarily separate; that is, an epipedon may overlap in whole or in part a subsurface diagnostic horizon. Those of both groups that occur in cold soils are described briefly next.

Mollic epipedons are relatively thick, dark mineral horizons that are rich in organic matter and nutrient elements, notably calcium and magnesium. Base saturation is 50% or more. The soil within the epipedon is well-granulated and friable when moist. Mollic epipedons are formed mainly by the incorporation of decomposed organic residues into the upper layers of the mineral soil. They contain at least 1% more organic matter than the underlying soil but, except in the case of some soils formed in volcanic ash, less than the amount of organic matter in histic epipedons, and they have dark gray or black colors when moist. In cold soils, most of which have a shallow solum over unaltered parent materials, a mollic epipedon is recognized only if the thickness of the dark material after mixing exceeds 18 cm or, in sands and stratified alluvium, 25 cm. If the solum thickness exceeds 75 cm, the minimum thickness of the mollic epipedon is 25 cm. Mixing may be assumed in the case of undisturbed soils with thinner dark upper horizons. Thus, a mollic epipedon can be recognized in a soil with a black upper mineral horizon only 5 or 6 cm thick if, after mixing with underlying horizons to a depth of 18 cm, the resulting color and organic matter content would satisfy the requirements for the horizon.

Umbric epipedons are identical to mollic epipedons except that base saturation is less than 50%. Most soils with umbric epipedons are strongly acidic.

Histic epipedons are highly organic surface horizons at least 20 cm thick that are continuously wet for at least 30 days each summer. They may consist entirely of organic matter or may be of mixed organic and mineral material. In mixed horizons the organic matter makes up at least 20% by weight, of the soil where no clay is present and at least 30% of the soil where the proportion of clay in the mineral fraction exceeds 60%. Proportional percentages of organic matter are required in soils with intermediate clay contents. The epipedon can be recognized with lower organic matter percentages in soils that have been deeply plowed. As much as 40 cm of volcanic ash or mineral alluvial deposits may overlie a histic epipedon.

Ochric epipedons have colors or organic matter percentages that do not meet the requirements for the other epipedons. In many cases this epipedon coexists with subsurface diagnostic horizons whose upper boundary approaches or even coincides with the surface of the mineral soil.

In contrast with the epipedons, which (except for the ochric epipedon) form by accumulation of organic matter at the surface, subsurface diagnostic horizons are the result of modification of the parent material either by chemical or physical alteration in place or by the migration and subsequent deposition of material carried in solution or suspension through the soil.

Cambic horizons are those that have been minimally altered. They have little or no increase in clay or organic matter attributable to illuviation and, in most cases, have lost mineral elements as a result of leaching. Alterations of the parent material may take the form of: (a) destruction of the original structure by plant roots, animal activity, or frost heaving; (b) development of soil, as distinct from parent material, structure; (c) weathering in place to form clay-size particles; (d) solution and removal or redistribution of carbonates; (e) chemical liberation of iron and aluminum from the parent material with a resultant change in soil color; or (f) segregation of iron oxides in wet soils to form red or brown mottles or streaks.

The *cambic horizon* exists in various forms, two of which are most common in cold soils. A *brown* cambic horizon occurs in well- or moderately well-drained soils in which the liberation of iron oxides results in more intense brown colors than in the unaltered parent material. Clay minerals may also have been synthesized in this kind of cambic horizon, especially in volcanic materials. A wet or *gleyed* cambic horizon occurs in soils with impeded drainage but with occasional periods of relative dryness. Free iron oxides are concentrated in mottles and concretions that form during periods when oxidizing conditions exist. The soil between the mottles retains the gray or bluish colors developed under reducing conditions during saturated periods. Horizons that are always saturated and that have uniform greenish or bluish colors inherited from the parent material are not recognized as cambic horizons.

No cambic horizon is permitted in sandy materials because of the difficulty in consistently distinguishing between sands that have undergone some alteration and those that have not. Some doubt exists as to the validity of this exclusion, particularly in the case of poorly drained sandy soils that are mottled and have obviously been subject to alternately reducing and oxidizing conditions. Horizons that still retain most of the original structure of the parent materials, including stratifications in alluvial sediments, are similarly not considered to be cambic horizons. It should be noted that horizons with fine stratifications developed as a result of freezing processes are not excluded on this basis.

Spodic horizons are illuvial layers in which organic matter, iron, and aluminum leached from overlying horizons have accumulated. They are formed by complex processes involving the combination of organic acids from decomposing plant litter at the soil surface with iron and aluminum in the upper part of the mineral soil, downward migration of these organomineral compounds in soluble form, and eventual precipitation. Typically, the precipitate in the upper part of the spodic horizon is highest in organic matter. Much of the iron and, to a somewhat greater extent, the aluminum is carried to greater depth before precipitation. As a result, the upper part of the spodic horizon is commonly black or dark reddish brown, and colors change with depth to reddish brown and yellowish brown.

In most cases the mineral layer from which the iron has been removed takes on a light gray color. This horizon, which is not by itself diagnostic at higher levels in the classification system, is called an *albic horizon*. Although albic horizons also occur in other situations, as above an argillic horizon, its presence is often helpful in field identification of a spodic horizon. Spodic horizons can exist, however, with no visible evidence of an albic horizon above them.

Also characteristic of spodic horizons are synthesized clay minerals with such weak or disordered crystal structure that they appear, on X-ray examination, to be amorphous. Each of these clay particles is surrounded by a thick shell of water. In the undisturbed condition soils containing these clays are fairly rigid, but under pressure or agitation the bond between the clay and its surrounding water is broken, and the soil becomes loose and watery. When the pressure is released or the agitation ceases, the original rigidity is restored. This property, called thixotropy, exists to some extent in all spodic horizons but is much less noticeable in sandy than in loamy materials. Similar, but not identical, clays exist in cambic horizons developed in fine volcanic ash. The two are difficult to distinguish in the field but can be identified by chemical tests.

Argillic horizons are enriched in fine clay that has been transported in suspension from overlying horizons. They are recognized either by an appreciable increase in clay percentage as compared with an upper horizon formed from the same parent material, by evidence of clay migration in the form of thin clay deposits (clay skins or cutans) on structural aggregates and in pores, or by clay bridges between sand grains. The horizon may be continuous or may occur as a series of discrete, roughly horizontal, clay lamellae. The horizon or the combined lamellae is required to be at least one-tenth as thick as the thickness of all overlying horizons; its minimum thickness is 15 cm in sands and 8 cm in loamy or clayey materials. Any epipedon or an albic horizon may occur above an argillic horizon.

Rarely, a horizon satisfies the requirements for both an argillic and a spodic horizon. Such horizons are considered to be spodic horizons.

Natric horizons are similar in most respects to argillic horizons but are high in sodium or magnesium and commonly exhibit a pronounced prismatic or columnar structure.

Calcic horizons are horizons of accumulation of calcium or magnesium carbonate. This accumulation commonly occurs below a mollic epipedon, a cambic horizon, or an argillic horizon, in the form of white powdery deposits or crusts on the lower surfaces of coarse fragments. In some soils the accumulation is within the mollic epipedon or argillic horizon and may even extend to the soil surface. In cold areas calcic horizons generally form in soils developed in calcareous materials or other parent materials that are high in bases. They are related only peripherally to major soil-forming processes.

Placic horizons are very thin, brittle pans cemented by iron, manganese, or an iron–organic matter complex. Commonly the pan is a single undulating sheet of indurated material that contains two layers with different cementing agents, but a series of these pans occurs in some soils. Placic horizons form only in areas with very high rainfall. In cold areas they occur mostly in association with spodic horizons, where the cementing agents are iron–organic complexes and iron oxides, and at the base of thick accumulations of peat in upland positions, where iron and manganese oxides are the principal components.

Fragipans are very firm, compact horizons that are hard when dry and brittle when moist. If shattered, most of them break into angular prismatic or blocky fragments. They are only slowly permeable to water, and on slopes water moving laterally above the fragipan may result in a light gray albic horizon at its surface. The apparent cause of the firmness of most fragipans is the development of fine bridges of nonswelling clay between coarse particles, but precipitated aluminum and iron oxides may be contributing factors especially in fragipans that underlie spodic horizons.

The several diagnostic horizons and their modes of development are described more fully in succeeding chapters. The mollic epipedon is discussed in Chapter 7, "Mollisols"; the umbric epipedon in Chapter 10, "Cryumbrepts"; the brown cambic horizon in Chapters 9, "Cryochrepts" and 8, "Cryandepts"; the gleyed cambic horizon in Chapter 11, "Cryaquepts"; the argillic and natric horizons in Chapter 6, "Alfisols"; and the spodic horizon, albic horizon, and fragipan in Chapter 5, "Spodosols." The placic horizon is discussed in Chapter 5 and in Chapter 12, "Histosols."

III. Horizon Designators

In addition to detailed written descriptions, horizons within pedons are identified by symbols—letters and numerals—that indicate genetic relationships among them. Capital letters, representing master horizons, commonly coincide with diagnostic horizons except in cases where the master horizons do not satisfy thickness requirements for the diagnostic horizons. The more commonly used master horizon designations will be described briefly here; more detailed descriptions of these and other master horizons and of the many subhorizon designations are given in *Soil Taxonomy* (Soil Survey Staff, 1975).

O horizons consist of organic soil materials. They may be made up entirely of organic litter at the surface of mineral soils or may be horizons in organic soils. If wet much of the summer and if of sufficient thickness, an O horizon can constitute a histic epipedon.

A horizons are horizons at or near the surface in which organic matter has been incorporated into the mineral soil. If thick enough the A horizon may be identical with a mollic or umbric epipedon. Horizons in which organic matter has accumulated as a result of illuviation, as in a spodic horizon, are not considered to be A horizons. Gray eluviated horizons such as the albic horizon, formerly labeled A2 horizons, are now called E horizons.

B horizons are altered subsurface horizons. This is the designation used for cambic, argillic, and spodic horizons, among others. Lower case letters are normally added to indicate more specifically the kind of modification that has taken place.

C horizons are horizons that are relatively unaffected by soil-forming processes related to organic activity. They may be completely unaltered or may be horizons with accumulations of carbonates or other soluble salts leached from above. Some have been compacted or cemented by silica or other leached material. Some calcic horizons and fragipans are C horizons.

IV. Nomenclature

A unique and controversial feature of the United States Taxonomy is its coined nomenclature. A completely new terminology rather than a redefinition of existing soil names was deemed to be necessary because of the great diversity of names—several for the same soils or specific characteristics in some instances and the same term for quite different soils or properties in others—in common use, and the large number of color terms and other poorly defined words from ordinary speech that serve as soil names in many countries (Smith, 1963). In addition to names for the diagnostic horizons, names have been coined for soil orders; many of these are based on the diagnostic horizons that serve as the principal distinguishing characteristic. Terms for characteristics that are the basis for classification in lower categories and a system for naming the classes in those categories also have been developed. The new nomenclature, as well as the taxonomy itself, has had only limited official acceptance outside of the United States, but unofficial recognition of its logic and utility, along with that of the classification system itself, is evidenced by its increasing use in published material by pedologists of other nations—except, notably, in the U.S.S.R. (Cline, 1980).

V. Categories of the Classification

The United States Taxonomy groups soils at six levels, or categories. The broadest category is the order. Lower categories, in which classes are defined successively more narrowly, are the suborder, the great group, the subgroup, the family, and the soil series. Few families or series have more than local or, at most, regional significance, and their distinguishing characteristics are related primarily to differences in parent materials. Virtually all properties associated with the principal soil-forming processes and major climatic and vegetative influences are accounted for at the subgroup and higher levels.

A. SOIL ORDERS

At the present time 10 soil orders have been defined. Soils representing six of them are recognized in cold areas; a seventh—the Aridisols—also is represented but no separate recognition is given at present to cold soils in that order. The six orders are:

1. *Entisols:* mineral soils with no diagnostic horizons other than an ochric epipedon. Soils with desertic characteristics, classified as Aridisols elsewhere, are included with Entisols in polar regions.
2. *Inceptisols:* mineral soils with no spodic or argillic horizon and no mollic epipedon unless base saturation below the epipedon is less than 50% or the soil is formed in volcanic materials. Most Inceptisols have a cambic horizon, but some are classified as Inceptisols only because of the presence of an umbric epipedon or of ice-rich permafrost and continuous wetness.
3. *Mollisols:* mineral soils with a mollic epipedon, except as noted under Inceptisols.
4. *Alfisols:* mineral soils with an argillic or natric horizon but no mollic epipedon. Base saturation in the argillic horizon is 35% or higher.
5. *Spodosols:* mineral soils with a spodic horizon or, if none, a placic horizon with the chemical characteristics of a spodic horizon that overlies a fragipan.
6. *Histosols:* Soils in which at least 40 cm of the upper 80 cm of the soil consists of organic soil material (60 cm of the upper 80 cm if the organic material is fibrous moss peat); lesser thicknesses are permitted if the organic material directly overlies bedrock or fragmental material.

In the nomenclatural system each order is assigned a descriptive syllable that is used as a formative element in developing the names of classes in lower categories. The formative elements for the orders that occur in cold areas are Entisols: *ent;* Inceptisols: *ept;* Mollisols: *oll;* Alfisols: *alf;* Spodosols: *od;* and Histosols: *ist.*

B. SOIL SUBORDERS

Names of classes in the two categories below the order are formed by prefixing other syllables that suggest properties that are emphasized in the taxonomy. The formative elements for suborders that occur in cold areas and the properties they represent are:

alb: albic horizon
and: volcanic materials
aqu: poor drainage; wetness
bor: cool soil temperatures
ferr: high iron content
fibr: relatively undecomposed organic materials (Histosols only)
fluv: alluvial deposits
fol: well-drained plant litter (Histosols only)
hem: partially decomposed organic materials (Histosols only)
hum: high organic matter content
ochr: brown cambic horizon
orth: central concept of order
psamm: sandy
sapr: highly decomposed organic materials (Histosols only)
umbr: ubric epipedon

Thus, for example, poorly drained or gleyed Inceptisols are classified as Aquepts; Mollisols with low soil temperatures, as Borolls; and Spodosols with an exceptionally high proportion of organic matter with respect to iron, as Humods. These separations are actually made on the basis of quantitative criteria that are given in *Soil Taxonomy* rather than on the general concepts just noted. The criteria are in some cases quite detailed.

C. GREAT GROUPS

A third syllable is added to form the names of great groups. In cold areas this is generally *cry,* indicating low soil temperatures—for example, Cryaquepts or Cryohumods—but where another property is of overriding importance it may take precedence—for example, Fragiquods for poorly drained Spodosols with a fragipan. In such cases the temperature designator is used at the subgroup level. Formative elements for names of great groups that occur in cold areas are:

bor: cool soil temperatures (Histosols only)
cry: cold soil temperatures
frag: fragipan
plac: placic horizon
sider: high iron content (Aquods only)
sphagn: sphagnum peat (Histosols only)

D. SUBGROUPS

Subgroup names are great group names modified by one or more adjectives. A *Typic* subgroup represents the central concept, but not necessarily the most extensive soils, of its great group. Other subgroups differ from the Typic in one or more respects, as indicated by the modifying adjective or adjectives. If they have properties approaching those of another great group, suborder, or order they are called *intergrade* subgroups. Those that have a property that is beyond the range of the Typic subgroup but is not used to define any other class in the system are called *extragrade* subgroups. Adjectives derived from the formative elements of other classes are used to name intergrades—for example, Histic Cryaquepts or Boralfic Cryorthods. Adjectives used for extragrade subgroups of cold soils are:

Aeric: better aerated (Aquic suborders only)
Hydric: free water (Histosols only)
Limnic: limnic layers (Histosols only)
Lithic: shallow bedrock or other material impermeable to roots
Pergelic: mean annual soil temperature of 0° C or less; permafrost
Ruptic: intermittent or broken diagnostic horizons within the pedon
Terric: mineral substratum (Histosol only)
Thapto: buried thick organic layer

Several advantages of this system of nomenclature are apparent, especially to those whose interest extends beyond national boundaries. Questions of priority in naming and of national sensibilities do not arise because names and their formative elements are derived primarily from classical languages rather than by a haphazard selection from differing folk usages. The formative elements are connotative. Pedologists familiar with their meanings can infer many of the important properties of a soil from its name alone and so can determine genetic relationships among soils of different geographic areas through the nomenclature. The names are reasonably short and the formative elements have been carefully selected so that, despite some protestations to the contrary (Orlovskiy, 1967), they are easily remembered and readily pronounced (Heller, 1963).

In succeeding chapters the dominant soil-forming processes in each of the orders that occur in cold regions are examined and the classification into suborders, great groups, and subgroups is discussed. Other major classification systems of importance in cold regions are described and compared with the United States Taxonomy in Chapter 13.

References

Baldwin, M., Kellogg, C. E., and Thorp, J. (1938). Soil classification. *U. S. Dept. Agr. Yearbook 1938*, 979–1001.

Cline, M. G. (1980). Experience with Soil Taxonomy of the United States. *Adv. Agron.* **33**, 193–226.

Heller, J. L. (1963). The nomenclature of soils, or what's in a name. *Soil Sci. Soc. Am. Proc.* **27**, 216–220.

Johnson, W. M. (1963). The pedon and the polypedon. *Soil Sci. Soc. Am. Proc.* **27**, 212–215.

Kellogg, C. E. (1963). Why a new system of soil classification? *Soil Sci.* **96**, 1–5.

Orlovskiy, N. V. (1967). Superposition of soil processes in relation to some problems in the classification and nomenclature of Siberian soils. *Sov. Soil Sci.,* 864–873.

Smith, G. D. (1963). Objectives and basic assumptions of the new classification system. *Soil Sci.* **96**, 6–16.

Soil Survey Staff. (1975). "Soil Taxonomy." U. S. Dept. Agr. Handbook 436, Washington, D.C.

Chapter 4

ENTISOLS

Entisols are soils that have been little altered by soil-forming processes. Many of them are young soils in which insufficient time has elapsed since deposition or exposure of the parent materials for genetic horizons to develop. Others are on steep slopes where erosion removes the products of rock weathering as quickly as they are formed or are in areas where soil turbation is so great that no expression of genetic processes is possible. Some are very old soils of cold, dry climates where chemical weathering of the parent materials is extremely slow. Entisols range from excessively drained soils, as in active sand dunes, to very poorly drained soils, as in marshes where the water table is constantly above the soil surface. The feature common to all but the wettest of them is the absence of any diagnostic horizon other than an ochric epipedon or, in sites that have a history of human occupation, an epipedon that is high in organic matter solely as a result of man's activities (an *anthropic* epipedon). In marshy areas Entisols may have a histic epipedon that consists entirely of organic materials over mineral soil in which no diagnostic horizon can be recognized.

Because of the relatively low intensity of mineral weathering in cold regions, Entisols occupy a higher proportion of the area than in most other regions. Four great groups of Entisols occur: *Cryofluvents* in well-drained portions of floodplains; *Cryorthents* in most upland positions; *Cryopsamments* in sandy parent materials; and *Cryaquents* in low-lying positions.

57

I. Cryofluvents

Deep, well-drained soils of floodplains bordering major rivers are of great importance in cold regions, especially in the boreal climatic belt. They support much of the agriculture of the regions, and major cities as well as small settlements are located on them. Most of these soils consist of stratified sandy and silty sediments. Clayey deposits are confined to the outer edges of floodplains, where floodwaters trapped behind natural levees along the main river channels move more slowly. Soils formed in clayey or other backwater deposits are seldom well-drained.

Most floods take place in the spring when snow in the drainage area melts rapidly and thick ice in the river channels impedes the free flow of water. Stratification in the alluvial sediments is commonly complex because of eddying around locally deposited ice floes. The strata vary in thickness and texture over short distances and are not always horizontal. Soils bordering rivers that originate in glaciers or ice fields are commonly underlain by thick deposits of gravelly sand. Depth to the gravelly substratum generally increases with increasing distance from the glacier.

The nature of well-drained soils of floodplains is less closely tied to climatic parameters than that of upland soils. In the uplands a recurring annual hydrothermal rhythm, with only minor variations from year to year, eventually results in the development of genetic layers that can be readily identified as either eluvial or illuvial horizons. Except where erosion is severe or there is continuing deposition of windborne material, Entisols in uplands gradually develop characteristics of one of the other orders. This is seldom the case in soils of active floodplains that are not subject to long periods of saturation. Floods do not occur every year, and the duration of floods when they do occur is variable. The boundaries and thickness of zones in the soil that lose or gain constituents differ from year to year. In cold continental areas with low precipitation, the soils may dry to the wilting point after 2 or 3 years with no floods; in other years, after heavy flooding, they may remain close to field capacity throughout the summer. The depth of thaw in soils with permafrost may be 40–60 cm greater in a year with a spring flood than in a following floodless year. The absence of a repeating pattern of movement of soil constituents, as well as the addition of fresh material during floods, is responsible for the failure of these soils to develop genetic horizons even over long periods of time (Konorovskiy, 1976).

In cold regions many well-drained soils of floodplains have brown or reddish brown streaks and patches, especially in a zone well above the permafrost table or, in soils with no permafrost, the midsummer water table. In contrast to poorly drained soils, the floodwaters that periodically saturate these soils probably are not deficient in dissolved oxygen. Since mottles are normally an indication of impeded drainage and periods of reduced oxygen supply, their presence in horizons that drain freely immediately after complete thawing is an anomaly that can cause confusion in classification. Two explanations for the mottling are possible. The first is that it is residual from a time when the soils were in a lower position with respect to the river level and were saturated most of the summer. The second and more likely explanation relates to the stratified nature of alluvial soils. Sandy strata thaw more rapidly than silty lenses, and moisture in the sand containing dissolved iron and other minerals migrates to the still frozen silty strata. During thawing of the silts, water tends to perch above textural boundaries and to migrate to the remaining patches of frozen material. This creates zones of maximum concentration of iron, which precipitates as the moisture is gradually lost to evapotranspiration.

In time, as rivers change course and are incised more deeply, the position of a soil with respect to the river channel changes. Flooding becomes less frequent, and the vegetation progresses from the willows and alder characteristic of low floodplains to typical upland plants of the region. This can result in significant changes in soil temperature. In interior Alaska, for example, the succession is from willows to balsam poplar, paper birch, white spruce, and eventually to black spruce forest with a sphagnum moss ground cover (Viereck, 1970). In central Siberia the succession leads to a forest of cedar, fir, and spruce (Aref'yeva, 1977). In tundra areas sedge tussocks, low-growing shrubs, and mosses become dominant. Under a cold continental climate, especially, the denser vegetation and the mat of organic matter that forms below it can result in the development of permafrost, increasing shallowness of the permafrost table, and, eventually, in soils that are constantly saturated by water perched above the perennially frozen substratum.

A. CLASSIFICATION OF CRYOFLUVENTS

Unfortunately, the classification of well-drained, cold Entisols of floodplains in the United States Taxonomy is not entirely satisfactory. Those of the warmer parts of the boreal belt that have no perennially frozen substratum are classified as *Typic Cryofluvents,* a placement that indicates clearly that the soils are subject to flooding, but those with permafrost (usually, in the boreal belt, at depths greater than 2 m) are classified as *Pergelic Cryorthents.* This is necessary because a principal criterion for the identification of Fluvents—an irregular decrease in organic matter content with depth—would apply as well to soils that are never flooded but that have been subject to cryoturbation and transport of fragments of surface organic matter to the underlying mineral soils. As a consequence, however, it is impossible to distinguish the well-drained soils of floodplains from well-drained upland Orthents except at the lowest level in the classification system, the series. It seems likely that the problem could be resolved by recognition of a moisture regime that is peculiar to flood-plains—brief periods of total saturation as a result of floods, followed by long periods during which the soil is at or below field capacity—and by using that moisture regime in the definition of Fluvents.

II. Cryorthents

The great group of Cryorthents includes well-drained upland soils in geologically young parent materials and soils that exist in an environment that inhibits the development of genetic horizons. They occur, mostly, in areas where loess blown from braided stream channels or outwash plains is deposited annually in quantities sufficient to overcome the effects of soil-forming pro-cesses, in areas such as steep slopes on mountains where natural erosion con-stantly exposes fresh soil material and in areas that are so cold that the vegetative cover is very sparse or absent. Well-drained soils formed in very gravelly material that is too coarse for recognition of a cambic horizon are also classified as Cryorthents, but sandy soils are classified separately as Cryopsam-ments. Most Cryorthents are classified as *Typic Cryorthents* or, where mean an-nual soil temperature is 0° C or lower and there is permafrost at a depth great enough that it does not affect surface processes, *Pergelic Cryorthents.*

A. LOESSIAL CRYORTHENTS

Strong winds blowing across unvegetated areas remove large quantities of surface materials and carry them considerable distances before depositing them as loess, usually in areas where the vegetation is sufficiently dense to hold it in place. The coarsest material is deposited first, and the finest material is carried the greatest distance. A common sequence of soil textures in areas that have ex-perienced deposition of windblown deposits is a narrow belt of sand dunes ad-joining the source area, a wider zone of very fine sandy soils beyond the dunes, and a broad belt of increasingly finer soils at greater distances. The thickness of the loess deposit decreases with increasing distance from the source. Major source areas of the loess are outwash plains, braided channels of rivers carrying glacial outwash materials, and barren ground or debris-laden, stagnant ice left behind by retreating glaciers. Maximum loess deposition in northern areas took place during periods of maximum glacial advance and of rapid melting of glaciers at the close of the Pleistocene era. Many large basins at the base of northern mountains, probably former outwash plains that are now fully vegetated, served as source areas for loessial material that now mantles the uplands surrounding them. In many cases these basins also have a surface mantle of loess derived from later source areas, including the rivers that now traverse them.

Except in areas that are still adjacent to retreating valley glaciers or to a continental glacier, as in Greenland, loess accumulation at the present time is largely confined, in cold regions, to narrow strips of land bordering braided rivers that carry glacial outflow water. Close to the rivers, where the average annual increment of freshly deposited loess is greatest, the continuing accretion of essentially unweathered material is large enough to offset the effects of soil-forming processes and to inhibit the formation of genetic horizons. Farther from the rivers, where only small amounts of loess are deposited each year, soil development that is characteristic of the region proceeds normally (Rieger and Juve, 1961).

Although soils with continuing loess accumulation have no diagnostic horizons other than an ochric epipedon, many of them in noncalcareous materials exhibit brown and grayish brown patches and streaks that contrast strongly with the generally gray soil material. This color variegation, which is similar to that observed in well-drained soils of floodplains, probably is the result partly of oxidation in place of the loess and partly of migration of iron into poorly defined mottles during periods of saturation in the course of the spring thaw. Because the soils are never saturated with oxygen-free water, the mottling is not the result of a gleying process but instead is considered to be the incipient phase of formation of a brown cambic horizon.

B. STEEPLY SLOPING CRYORTHENTS

Strongly sloping soils of mountains, especially those that are sparsely vegetated, may have no diagnostic horizons simply because most products of weathering are lost to erosion as quickly as they are formed. Soils of this kind are invariably shallow and stony and are unlikely to develop genetic horizons as long as erosion continues. On less steeply sloping land, many of these primitive soils are in the initial stages of development and, in time, will have diagnostic horizons characteristic of soil orders other than Entisols. The first stage of development occurs when lichens, which are capable of penetrating rocks and decomposing them biochemically, become established. This generally results in the formation of a thin, dark surface horizon that is strongly acidic and higher in clay and sesquioxides than the parent rock (Rozhnova and Schastnaia, 1959) and later in colonization by higher plants. Organic accumulation eventually increases to the point that soil-forming processes can proceed. Development may be delayed, however, in parent materials that are highly resistant to weathering. In the mountains of southeastern Siberia, for example, soils with the typical horizon sequence of Spodosols have formed in granitic parent material under a larch forest, whereas soils under the same vegetation in denser amphibolite have not yet developed these horizons although they do have some of the chemical characteristics of Spodosols (Sokolova, 1964; Sokolova and Smirnova, 1965).

C. CRYOTURBATED CRYORTHENTS

Severe cryoturbation in coarse-grained soils may prevent establishment of a stable vegetative cover and the formation of genetic horizons. These soils commonly have stony or gravelly surface horizons. In microdepressions, as in troughs between polygons, enough fine material and moisture may accumulate to support a permanent vegetation. Eventually, as organic matter penetrates the soil under the vegetation, an intermittent umbric or mollic epipedon may develop. Soils with this kind of complex development are no longer considered to be Entisols, but are members of ruptic subgroups such as *Ruptic–Entic Cryumbrepts* or *Ruptic–Entic Cryoborolls*. Soils in which cryoturbation has been severe in the past but that are now more stable because of an amelioration of the climate, as in the mountains of southern British Columbia, still retain the surface and internal signs of disturbance but have a permanent cover of lichens and other plants and incipient forms of genetic horizons (Van Ryswyk, 1971).

D. VERY GRAVELLY CRYORTHENTS

Soils formed in very coarse materials in which no cambic or other diagnostic horizon can be recognized are classified as Cryorthents even where horizons with brown colors and, in some cases, horizons of calcium carbonate accumulation have developed under a permanent vegetation. Apart from texture, these soils are identical with Cryochrepts in materials that contain a higher proportion of fines. Typical sites for such soils are old beach ridges, including some that are as much as 100 km from the present shoreline (Mills and Veldhuis, 1978). Some very coarse soils, especially those formed from shattered limestone or other dense rock, have a surface horizon of frost-rived gravel from which all fine material has been removed by wind and, at a depth of 5–10 cm, a layer containing appreciable quantities of silt (Holowaychuk *et al.*, 1966) (Fig. 15). Laterally growing roots from tundra plants in microdepressions penetrate this layer to supplement their moisture supply.

E. CLAYEY CRYORTHENTS

Most Cryorthents in tundra areas are coarse-grained, but there are exceptions. On bentonitic shales in northern Alaska, for example, dark clayey soils occupy narrow ridge tops and other sites with exceptionally good surface drainage (MacNamara and Tedrow, 1966). These soils become quite dry in the summer and crack to depths as great as 20 cm. Sodium sulfate and other salts

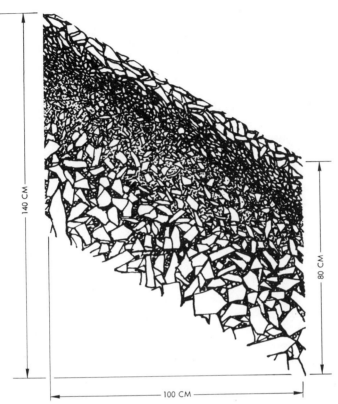

FIGURE 15. *Vertical distribution of fine and coarse material in barren, stony soil on a steep slope in northwestern Alaska. Frost-rived gravel free of fines makes up the surface layer; a zone of maximum silt content occurs in the 13–23 cm section (Holowaychuk et al., 1966). (Reprinted by permission of the National Technical Information Service, U. S. Department of Commerce, Springfield, Va.)*

may accumulate on the surface. When moisture from late summer precipitation flows into the cracks and migrates into the intervening soil, swelling of the clay creates an undulating surface relief resembling the gilgai relief that is commonly associated with soils of the order of Vertisols (Soil Survey Staff, 1975). The bentonitic soils of the arctic, however, do not have the slickensides or wedge-shaped structure that is characteristic of Vertisols and are classified as Cryorthents.

F. CRYORTHENTS OF VERY COLD AREAS

Cryorthents are the dominant soils of the high arctic, where temperatures and precipitation rates are very low. Higher vegetation covers less than 25% of the surface and is confined largely to troughs between polygons and to swales and depressions where drifting snow accumulates and seepage water keeps the soil moist during the short summer. Although well-drained soils in some areas of this "polar desert" are frequently moistened by light rain or snow (Tarnocai, 1978), most of them have many characteristics in common with soils of warm deserts. Coarse, angular pebbles cover the surface; soil pH is high, and the soil solution has a high electrolyte content; carbonates coat the undersides of pebbles at shallow depth in soils formed in calcareous materials; iron oxide similar to desert varnish remains after carbonates are leached out and forms a thin veneer on pebbles and stones near the surface; and salt crusts are common (Targul'yan, 1959; McMillan, 1960; Tedrow and Douglas, 1964; Tedrow, 1966; Bunting and Hathout, 1971; Cruickshank, 1971). Soils formed in coarse grantic materials are acidic and low in salts but are otherwise comparable to the basic soils (Evans and Cameron, 1979). The principal points of difference between these and the warm desert soils are their low temperatures and the presence of dry or ice-rich permafrost at depths ranging from 1 m in very gravelly sands to 30 cm in soils with clayey substrata. Although the soils give little evidence of organic matter accumulation, algae that grow abundantly on the surface may contribute colorless organic matter that is disseminated throughout the soil by cryoturbic processes (Tedrow, 1966).

Extreme desert conditions prevail in the dry valleys of Antarctica, where the climate is too cold and dry to support higher vegetation. Soils here contain virtually no organic matter except in nesting areas where guano accumulates on the surface and in a few areas where lichens, mosses, and algae survive (Tedrow and Ugolini, 1966). There is active physical weathering, but the rate of chemical weathering is very low. Most soils are high in soluble salts. Although the soil moisture content is low and there is little frost stirring, a polygonal surface relief is common as a result of the growth of sand wedges in very dry areas and ice wedges in areas with somewhat higher precipitation (Berg and Black, 1966). Ice-rich permafrost underlies most soils at depths ranging from a few centimeters to 2 m (Claridge and Campbell, 1977).

An outstanding characteristic of soils of Antarctica is their high concentration of soluble chlorides, nitrates, and sulfates. It seems likely that most of the salts, especially the chlorides, originated in the ocean and were added to the soils by precipitation (Campbell and Claridge, 1969; Pastor and Bockheim, 1980), but at least some of them were released by chemical weathering of the parent materials (Everett, 1971). Despite the low moisture content of the soils, the salts migrate upwards in winter, in response to very low temperatures at the soil surface, in extremely thin films of adsorbed water on mineral grains. This movement is not balanced by downward migration toward the permafrost in summer because of evaporation at the surface (Ugolini and Anderson, 1973). The net result in older soils of the driest areas is an indurated, highly saline horizon as much as 15 cm thick, just below the pebbly or bouldery surface layer from which all fines have been winnowed by winds (Everett, 1971). In younger soils and in soils of areas with somewhat higher precipitation, the salts accumulate on the undersides of stones or are fairly uniformly disseminated throughout the soil profile (Claridge and Campbell, 1977). Carbonates accumulate beneath stones in soils formed in calcareous materials. In depressions and basins salt crusts several centimeters thick occur on the soil surface. In some areas the salts accumulate in saline lakes (Tedrow and Ugolini, 1966).

Chemical weathering, although extremely slow, affects these very cold soils. An iron–manganese crust similar to desert varnish coats pebbles and rocks exposed at the surface. The upper 20–30 cm of older soils, some of which have had no glacial cover for several million years, have yellowish brown colors. The thickness of the "solum" increases with age of the soils (Bockheim, 1980). Both physical disintegration and chemical weathering decrease with depth; below 30 cm pebbles are angular and have fresh surfaces. The degree of both physical and chemical weathering is closely related to age. The older and more highly weathered soils have a higher proportion of silt and clay, especially in salt-indurated horizons (Campbell and Claridge, 1969; Everett, 1971). Although any clay transformations are extremely slow (Bockheim, 1980), a somewhat higher proportion of montmorillonite in the clay fraction of the more highly weathered horizons than in the younger soils has been reported (Everett, 1971). The dominant clay mineral weathering process, however, is the slow hydration of micas (Pastor and Bockheim, 1980). It is possible that much of this weathering took place under a warmer, more humid climate (Jackson *et al.,* 1977), but there is evidence to indicate that weathering processes are continuing in the present environment (Ugolini and Anderson, 1973).

Most well-drained soils of the high arctic presently are classified as *Pergelic Cryorthents* in the United States Soil Taxonomy. Although many are dry and salty and meet the requirements for Aridisols (other than temperature), no provision is made for soils with a pergelic temperature regime in that order. Because soil is defined in the Taxonomy as earthy material that is capable of supporting higher vegetation, no soil can be recognized in areas like Antarctica, where extreme cold makes it impossible for higher plants to survive. It is apparent, however, that soil-forming processes similar to those in other desert areas do go on there. Furthermore, the "soils" are not entirely lifeless; although smaller in number than in most soils, a wide variety of microorganisms, including bacteria, algae, fungi, and protozoa are present (Cameron *et al.,* 1970; Jordan *et al.,* 1978) and undoubtedly contribute to weathering processes just as in warmer areas. As the Taxonomy develops and is revised, it would seem desirable to consider a more inclusive definition of soil and to extend the temperature range of Aridisols to pergelic soils.

G. OTHER CRYORTHENTS

Although most Cryorthents are classified in the Typic or Pergelic subgroups, some have properties that are important enough to warrant separate recognition. Soils with a layer of fresh volcanic ash at or near the surface in which no diagnostic horizons have yet developed are separated in a subgroup of *Andeptic Cryorthents* because it is likely that, in a fairly short time, they will acquire the characteristics of Cryandepts (Chapter 8). Some such soils under coniferous forest have thin layers that are high in clay and that probably are precursors of an argillic horizon. These soils are classified as *Alfic Andeptic Cryorthents*. Entisols that are occasionally saturated and have some mottles are classified as *Aquic Cryorthents*. These soils occur for the most part at high elevations close to tree line. Soils with shallow bedrock, regardless of temperature, are *Lithic Cryorthents*.

III. Cryopsamments

Sand dunes border outwash plains and braided river channels and occur in strips of varying width along coasts. In some cases they are still active and support little or no vegetation, but in many places they are now stabilized and support vegetation ranging from forest to tundra or beach grasses (Rieger *et al.*, 1979). Except where a layer of silty loess more than 25 cm thick has been deposited on the surface or where there has been sufficient time since stabilization for diagnostic horizons other than an ochric epipedon to develop, soils on the dunes are classified as *Typic Cryopsamments* where there is no permafrost and *Pergelic Cryopsamments* where mean annual soil temperatures are as low as 0° C. Active, completely unvegetated dunes are not considered to be soils in terms of the United States Taxonomy.

In most cold, forested areas, the Cryopsamments are relatively short-lived after stabilization of the dunes. Even in areas where the mean annual precipitation is as low as 200 mm, thin spodic horizons begin to form soon after stabilization, and the soils eventually become Spodosols. Those with horizons that are obviously transitional to spodic horizons are classified as *Spodic Cryopsamments*. In some very dry areas, brown colors coat the sand grains, but because no cambic horizon can be recognized in sandy materials, the classification of the soils does not change.

In tundra areas ice-rich permafrost eventually forms at some depth as the vegetation becomes established (Rickert and Tedrow, 1967; Everett and Parkinson, 1977). On high points of dunes the permafrost table may be several meters deep, and the soils eventually accumulate enough organic matter in the upper layers so that an umbric or mollic epipedon can be recognized. Albic and spodic horizons may develop after long periods of stability (Everett, 1979). On lower dunes the soils become saturated over a shallow permafrost table and acquire the characteristics of Aquepts. In a large area in the arctic coastal plain of Alaska, Pergelic Cryopsamments on low dunes are associated in a complex pattern with sandy Pergelic Cryaquepts (Chapter 11) and Histosols (Chapter 12) in swales and depressions (Rieger *et al.*, 1979; Everett, 1979).

Cryopsamments also occur in other relatively young sandy materials, including coarse-textured alluvium on terraces, delta remnants in areas of rapid glacial retreat such as Greenland (Ugolini, 1966), and detritus from coarse-grained rock (Ragg and Ball, 1964). Future development in most of these soils probably will be similar to that of soils on dunes. Some sandy mountain soils have a few thin lamellae of clay; these soils are apparently developing in the direction of Alfisols (Chapter 6) and are classified as *Alfic Cryopsamments.*

IV. Cryaquents

Poorly drained soils are extensive in cold areas, but they are classified as Entisols only if they have no permafrost and no diagnostic horizons other than an ochric epipedon or a histic epipedon consisting entirely of organic materials. For the most part, Cryaquents are restricted to floodplains, outwash plains at the mouths of existing glaciers, lake margins, and coastal marshes. Soils in these areas are either almost continually saturated, stratified, or too sandy to permit recognition of a cambic horizon.

Sandy Cryaquents occupy extensive areas at the terminus of glaciers that originate in high mountain ice fields. Glacial meltwater keeps the water table in these soils very close to the surface at all times when they are not frozen. During periods of rapid glacial melting, the soils are often flooded. They generally have uniform gray or bluish colors, but some are mottled. The vegetation consists mostly of willows, alder, other water-tolerant shrubs, sedges, grasses, and mosses. In some places trees like black spruce or lodgepole pine are dominant. Commonly there is a mat of mossy organic matter on the surface; this may become thick enough to convert the soils to Histosols (Chapter 12). Much the same situation prevails in frequently flooded areas bordering streams, but soils there may have lenses of fine material and commonly support forests of balsam poplar and other hydrophyllic trees.

Cryaquents in coastal marshes and deltas and on swampy margins of inland lakes and bogs may have silty or clayey textures. Because these soils are almost constantly saturated, all iron exists in the reduced form and imparts a bluish or greenish color to the soil. Little, if any, is lost from the soil by leaching. Typically, when samples of the soil are exposed to air, oxidation rapidly causes a change in color to gray or olive. These soils also support a water-loving vegetation and, like the sandy soils, may have a mat of decaying organic matter on the surface.

Only two subgroups of Cryaquepts are established, *Typic Cryaquents* for most of the soils and *Andaqueptic Cryaquents* for soils with at least an upper layer that is high in volcanic ash.

References

Aref'yeva, Z. N. (1977). Evolution of floodplain soils of the taiga zone (as illustrated by the Kul'-Yegan River floodplain). *Sov. Soil Sci.* **9,** 49–58.

Berg, T. E., and Black, R. F. (1966). Preliminary measurements of growth of nonsorted polygons, Victoria Land, Antarctica. *In* "Antarctic Soils and Soil Forming Processes" (J. C. F. Tedrow, ed.), pp. 61–108. Am. Geophys. Un. Pub. 1418. Washington, D.C.

Bockheim, J. G. (1980). Properties and classification of some desert soils in coarse-textured glacial drift in the Arctic and Antarctic. *Geoderma* **24,** 45–69.

Bunting, B. T., and Hathout, S. A. (1971). Physical characteristics and chemical properties of some high-arctic organic materials from southwest Devon Island, Northwest Territories, Canada. *Soil Sci.* **112,** 107–115.

Cameron, R. E., King, J., and David, C. N. (1970). Soil microbial ecology of Wheeler Valley, Antarctica. *Soil Sci.* **109,** 110–120.

Campbell, I. B., and Claridge, G. G. C. (1969). A classification of frigic soils—the zonal soils of the Antarctic continent. *Soil Sci.* **107,** 75–85.

Claridge, G. G. C., and Campbell, I. B. (1977). The salts in Antarctic soils, their distribution and relationship to soil processes. *Soil Sci.* **123,** 377–384.

Cruickshank, J. G. (1971). Soils and terrain around Resolute, Cornwallis Island. *Arctic* **24,** 195–209.

Evans, L. J., and Cameron, B. H. (1979). A chronosequence of soils developed from granitic morainal material, Baffin Island, N. W. T. *Can. J. Soil Sci.* **59,** 203–210.

Everett, K. R. (1971). Soils of the Meserve Glacier area, Wright Valley, South Victoria Land, Antarctica. *Soil Sci.* **112,** 425–438.

Everett, K. R. (1979). Evolution of the soil landscape in the sand region of the arctic coastal plain as exemplified at Atkasook, Alaska. *Arctic* **32,** 207–223.

Everett, K. R., and Parkinson, R. J. (1977). Soils and landform associations, Prudhoe Bay area, Alaska. *Arc. Alp. Res.* **9,** 1–19.

Holowaychuk, N., Petro, J. H., Finney, H. R., Farnham, R. S., and Gersper, P. L. (1966). Soils of Ogotoruk Creek watershed. *In* "Environment of the Cape Thompson Region, Alaska" (N. J. Wilimovsky, ed.), pp. 221–273. U. S. Atomic Energy Comm., Div. Tech. Inf. Ext., Oak Ridge, Tennessee.

Jackson, M. L., Lee, S. Y., Ugolini, F. C., and Helmke, P. A. (1977). Age and uranium content of soil micas from Antarctica by the fission particle track replica method. *Soil Sci.* **123,** 241–248.

Jordan, D. C., Marshall, M. R., and McNicol, P. J. (1978). Microbiological features of terrestrial sites on the Devon Island lowland, Canadian Arctic. *Can. J. Soil Sci.* **58,** 113–118.

Konorovskiy, A. K. (1976). Characteristics of formation and properties of the floodplain soils of central Yakutiya. *Sov. Soil Sci.* **8,** 125–132.

McMillan, N. J. (1960). Soils of the Queen Elizabeth Islands (Canadian Arctic). *J. Soil Sci.* **11,** 131–139.

MacNamara, E. E., and Tedrow, J. C. F. (1966). An arctic equivalent of the Grumusol. *Arctic* **19** 145–152.

Mills, G. F., and Veldhuis, H. (1978). A buried paleosol in the Hudson Bay lowland, Manitoba: Age and characteristics. *Can. J. Soil Sci.* **58**, 259–269.

Pastor, J., and Bockheim, J. G. (1980). Soil development on moraines of Taylor Glacier, lower Taylor Valley, Antarctica. *Soil Sci. Soc. Am. Jour.* **44**, 341–348.

Ragg, J. M., and Ball, D. F. (1964). Soils of the ultra-basic rocks of the Island of Rhum. *J. Soil Sci.* **15**, 124–133.

Rickert, D. A., and Tedrow, J. C. F. (1967). Pedologic investigations on some aeolian deposits of northern Alaska. *Soil Sci.* **104**, 250–262.

Rieger, S., and Juve, R. L. (1961). Soil development in recent loess in the Matanuska Valley, Alaska. *Soil Sci. Soc. Am. Proc.* **25**, 243–248.

Rieger, S., Schoephorster, D. B., and Furbush, C. E. (1979). "Exploratory Soil Survey of Alaska." U. S. Dept. Agr., Soil Cons. Serv. Washington, D.C.

Rozhnova, T. A., and Schastnaia, L. S. (1959). Study of the interrelationships of vegetation and soil in the Karelian Isthmus. *Sov. Soil Sci.,* 15–24.

Soil Survey Staff. (1975). "Soil Taxonomy." U. S. Dept. Agr. Handbook 436, Washington, D.C.

Sokolova, T. A. (1964). Effect of rocks on Podzol formation. *Sov. Soil Sci.,* 233–240.

Sokolova, T. A., and Smirnova, G. Ya. (1965). Development of podzolic soils on granite. *Sov. Soil Sci.,* 642–649.

Targul'yan, V. O. (1959). The first stages of weathering and soil formation on igneous rocks in the tundra and taiga zones. *Sov. Soil Sci.,* 1287–1296.

Tarnocai, C. (1978). Distribution of soils in northern Canada and parameters affecting their utilization. *Trans. 11th Int. Cong. Soil Sci.,* Vol. 3, 332–347.

Tedrow, J. C. F. (1966). Polar desert soils. *Soil Sci. Soc. Am. Proc.* **30**, 381–387.

Tedrow, J. C. F., and Douglas, L. A. (1964). Soil investigations on Banks Island. *Soil Sci.* **98**, 53–65.

Tedrow, J. C. F., and Ugolini, F. C. (1966). Antarctic soils. *In* "Antarctic Soils and Soil Forming Processes" (J. C. F. Tedrow, ed.), pp. 161–177. Am. Geophys. Un. Pub. 1418. Washington, D.C.

Ugolini, F. C. (1966). Soils of the Mesters Vig District, northeast Greenland: I. The Arctic Brown and related soils. *Medd. om Gron., Bd.* **176**, *No. 1,* 22 pp.

Ugolini, F. C., and Anderson, D. M. (1973). Ionic migration and weathering in frozen Antarctic soils. *Soil Sci.* **115**, 461–470.

Van Ryswyk, A. L. (1971). Radiocarbon date for a cryoturbated alpine Regosol in south central British Columbia. *Can J. Soil Sci.* **51**, 513–515.

Viereck, L. A. (1970). Forest succession and soil development adjacent to the Chena River in interior Alaska. *Arc. Alp. Res.* **2**, 1–26.

Chapter 5

SPODOSOLS

Spodosols, simply defined, are soils that contain a spodic horizon; that is, soils in which there has been translocation of organic matter, iron, and aluminum from the upper part of the soil to a lower horizon. Some soils with a placic horizon that overlies a fragipan, but with no spodic horizon, are also classified as Spodosols. Typically, Spodosols consist of an upper layer of decaying organic matter (the 0 horizon), a gray horizon of varying thickness (the albic horizon), and a black to reddish brown horizon that becomes yellower or less intense in color with depth (the spodic horizon).

Spodosols are most extensive under coniferous forests in areas with a cool, humid climate and under ericaceous alpine tundra vegetation, but they also occur under some mixed or deciduous forests, in coarse-textured materials in areas with a cold continental climate, in areas with grassy vegetation beyond the cold limit of trees, and in a few low-elevation tundra areas north of the tree line. Except in coastal areas with exceptionally high precipitation rates, they form primarily in noncalcareous sandy, coarse-loamy, or coarse-silty materials. In many places, however, Spodosols develop in materials from which clay or carbonates, or both, have been removed by intensive leaching (Stobbe and Wright, 1959; Heilman and Gass, 1974; Buurman *et al.*, 1976).

I. Genesis of Spodosols

A. CONDITIONS OF FORMATION

The principal requirements for the formation of Spodosols are: *(a)* permeable parent material; *(b)* a moisture regime in which there are few or no periods of dryness and in which excess moisture passes completely through the solum; *(c)* an acidic soil solution; *(d)* domination of the exchange complex by hydrogen and aluminum; and *(e)* continual addition to the soil solution of organic acids that can combine with iron and aluminum to form soluble organometallic complexes. In such an environment there is intense weathering of primary and secondary (clay) minerals and, generally, little illuviation of whole fine clay. In many soils with moderately high clay content in areas with high precipitation rates, however, clay illuviation dominates in an early stage of soil development and illuviation of organometallic complexes follows later (Fridland, 1958; Parfenova and Yarilova, 1960; Gorbunov, 1961; Targul'yan, 1964; Brydon, 1965; Laflamme *et al.,* 1973; DeKimpe and McKeague, 1974; Guillet *et al.,* 1975; Hole, 1975; DeKimpe, 1976). Soils in which this occurs are considered to be Spodosols if a horizon in which illuviation of organometallic complexes can be shown to be the dominant formation process develops in or above the horizon of clay accumulation.

B. RELATIONSHIP TO "PODZOLS"

Most Spodosols of cold regions were called Podzols before adoption of the new taxonomic system in the United States (the term is still used in other classification systems), but Podzols and podzolic groups in the earlier classification included many soils that are not now Spodosols. The early Russian view, as exemplified by Rode (1937), was that Podzols are formed by a complete breakdown of all minerals except quartz in the upper part of the soil and resynthesis of clay minerals from the decomposition products in a lower horizon. Migration of whole clay in suspension was considered to be unimportant. Along with this "podzolization" process, iron and aluminum were believed to have migrated to the lower horizon as ions in solution, either alone or in combination with organic acids, or as sols under the protection of organic acids or dissolved silica. The typical Podzol was considered to be one in which a bleached or podzolic horizon overlies a horizon enriched with clay (Muir, 1961). This concept is reflected in the current official soil classification of the U.S.S.R. (Rozov and Ivanova, 1967). Soils in which the principal accumulation products are organic matter, iron, and aluminum are relegated to a genus of iron–aluminum–humus Podzols. Many Soviet pedologists, however, recognize that whole clay movement does take place in soils and make a clear

distinction between the processes of whole clay migration and iron, aluminum, and organic matter migration (Fridland, 1958; Karpachevskiy, 1960;
Targul'yan, 1964). The difference between the two processes is widely
recognized in western Europe and elsewhere, and the term "podzolization"
has increasingly been reserved for the process that involves the complexing of
organic acids and metals (Pedro *et al.*, 1978). Nevertheless, because of the long
history of the term and its continuing application to both processes by some
pedologists, it seems better to forego its use and to refer to the sequence of
events involved in the formation of the spodic horizon as the spodic process.*

C. ORGANIC ACIDS

The spodic process involves: *(a)* the production of chemically active organic
acids in the 0 horizon; *(b)* the attack by these acids on silicate and other minerals
in the upper part of the soil; *(c)* a combination of the acids and the mineral
decomposition products; (*d*) a downward migration of the resulting organometallic complexes; and (*e*) the precipitation, at least in part, of the complexes to form the spodic horizon. Secondary processes, which may or may not
require the participation of organic acids, go on simultaneously with this basic
process.

Major end products of biological decomposition of organic materials in the 0
horizon are two groups of soluble organic acids—humic acid and fulvic
acid—and humin, an insoluble residue. Humin does not react with other soil
constituents and is not considered further. Both humic acid and fulvic acid are
actually made up of a number of related organic compounds and can be split
into several component fractions based on their affinity for different cations
and their activity in the soil (Dormaar, 1964; Lowe, 1975). Insofar as they are
involved in the development of soil genetic horizons, however, each group can
be treated as an individual entity. The groups are defined by their solubility in
acidified water. Both can be extracted from the soil by an alkaline aqueous solution; humic acid precipitates on acidification of the extract, but fulvic acid remains in solution.

It is likely that both humic acid and fulvic acid form as decay products in the
0 horizons of all soils of temperate and cold regions, but the proportion of fulvic
acid is greater in litter, like that under coniferous forest and ericaceous heath,

---

* "Podzol" is a Russian folk term that refers specifically to the albic horizon. It is derived from
the Russian words, "zola", meaning ash, and "pod", which translates literally as "underneath",
but which is also an idiomatic Russian term for soil (Muir, 1961). In developing the terminology
for the United States Taxonomy, the Greek "spodos", meaning wood ash, was selected as the root
word for the order of Spodosols because of its homonymic similarity to Podzol (Heller, 1963),
despite the fact that the meanings of "pod" and "spodos" are entirely different. The result is that
the name of the order and of the spodic horizon continue to refer to the albic horizon that commonly (but not invariably) overlies the definitive illuvial horizon.

that is high in aluminum and low in bases (Bel'chikova, 1966; Messenger, 1975). Many cold soils nevertheless have more humic acid than fulvic acid in the 0 horizon, although, as in the mountains of southern Siberia, fulvic acid is dominant in the organic matter of the coldest soils (Rubilin and Dzhumagulov, 1977). Because of its low solubility in an acidic medium, however, little humic acid is carried downward with percolating water in acidic soils (Alekseyeva and Pereverzev, 1974; Nikonov, 1979).

Although chemically similar, humic acid in cold soils is not as complex or as well-developed as that in warmer soils and has only low resistance to chemical degradation (Schnitzer and Vendette, 1975). There is evidence to indicate that at least part of the fulvic acid in 0 horizons is derived from the decomposition of humic acid. Fungi, which have greater tolerance for high aluminum concentrations than actinomycetes or bacteria, are dominant in the microflora of acidic litter and probably are a major factor in this alteration (Messenger, 1975; DeKimpe and Martel, 1976). The conversion of humic acid to fulvic acid results in a decrease in carbon, hydrogen, and nitrogen, an increase in oxygen, and increased acidity of the soil solution (Wright and Schnitzer, 1963). Fulvic acids have about 10% more oxygen and 5% less carbon than humic acids (Schnitzer and Skinner, 1968). Carbon makes up 44–50%, by weight, of the average fulvic acid molecule (Schnitzer, 1970). The acid in solution has an intense yellowish color (Messenger *et al.*, 1972), due mainly to its phenolic acid (a ring structure incorporating a hydroxyl group) component (Chen *et al.*, 1978).

An additional source of organic acids is the leachate, by rainwater, from leaves, needles, and tree bark (Muir *et al.*, 1964). This leaching is greatest from plants growing in nutrient-deficient soils (Schnitzer and DeLong, 1955). No iron, aluminum, manganese, or silicon is extracted from the forest canopy as rain passes through it, but on entering the 0 horizon, the organic compounds combine with those metals and their concentration in solution increases markedly (Singer *et al.*, 1978). Addition of the organic material causes a significant drop in the pH of the rainwater as it passes through the canopy and the 0 horizon (Ugolini *et al.*, 1977). The total amount of metal-complexing organic compounds contributed by canopy drip is low compared with that derived from the 0 horizon itself, but its importance is enhanced by its association with rainfall that percolates through the soil immediately (Malcolm and McCracken, 1968).

Fulvic acids have a wide range of molecular weights. An extract from the spodic horizon of a soil in eastern Canada contained fractions ranging in molecular weight from 175 to 3570 (Khan and Schnitzer, 1971), with a number average weight of about 670 (Schnitzer and Skinner, 1963) to 957 (Kodama and Schnitzer, 1977). In a Spodosol from the Cascades of Washington, it was reported to be 837–845 (Dawson *et al.*, 1978). The molecules are believed to be polymers made up either of isolated aromatic (benzene) rings linked by aliphatic chains (Hansen and Schnitzer, 1967) or of an aromatic core to which aliphatic chains are attached (Kodama and Schnitzer, 1970; Pollack *et al.*, 1971). About 60% of the average fulvic acid molecule consists of oxygen-containing functional groups—carboxyls (COOH), hydroxyls (OH), and, to a lesser extent, carbonyls (CO) (Wright and Schnitzer, 1963). The large number of these groups is of considerable importance because they are responsible, through hydrogen bonding with water, for the solubility of fulvic acid at low pH (MacCarthy and O'Cinneide, 1974; Goh and Reid, 1975), and they are the groups involved in the formation of organometallic complexes (Wright and Schnitzer, 1963; Anderson *et al.*, 1977).

As seen through an electron microscope fulvic acid at pH 2 to 3 consists of elongated fibers with an open structure. Above pH 4, in the reaction range of most Spodosols, the fibers mesh to form a spongelike structure. At pH 9 it forms sheets and at pH 10 it coalesces into fine grains (Chen and Schnitzer, 1976). The more open molecular arrangement at low pH is apparently better adapted to chemical activity in the soil. Humic acid goes through a similar series of transformations, but the major transitions occur at pH values higher than 6.

D. THE SPODIC PROCESS

Fulvic acid is by far the dominant organic fraction below the 0 horizon in Spodosols and is the major agent in the translocation of iron, aluminum, and manganese from the upper soil horizons to the spodic horizon (Schnitzer and Desjardins, 1962; Wright and Schnitzer, 1963; Hansen and Schnitzer, 1967; Archegova, 1968; Anderson *et al.*, 1977; McKeague and Sheldrick, 1977; Nikonov, 1979). The spodic process begins when fulvic acid in the leachate from the canopy and the upper part of the 0 horizon combines with metals from humus and primary and secondary minerals to form soluble organometallic complexes. The process cannot proceed, however, until most of the calcium and magnesium in the upper horizons has been removed by leaching, as the presence of even small amounts of these elements results in precipitation of the complexes (Stobbe and Wright, 1959; Wright and Schnitzer, 1963). The principal complexing agents are the carboxyl groups, which form an electrovalent bond with the metal (Schnitzer and Skinner, 1964). Also important are the phenolic hydroxyl groups, which probably incorporate the metal in the ring structure by a chelation process (McHardy *et al.*, 1974). The organometallic complexes may migrate downward in true aqueous solution or in the form of highly dispersed colloids (Gorbunov, 1961; Bloomfield *et al.*, 1976). Although fulvic acid may become complexed with any metal, its strongest affinity is for iron and, to a lesser extent, aluminum (Rozhnova and Schastnaia, 1959; Schnitzer, 1969; Schnitzer and Skinner, 1963, 1965; Schnitzer and Hansen,

1970). It can extract and bring into solution these elements from practically insoluble oxides (Schnitzer and Desjardins, 1969) and from primary and secondary minerals (Schnitzer and Kodama, 1976). At low pH, fulvic acid molecules can enter the interlamellar spaces of expansible clay minerals, but above pH 5 or 5.5, it is adsorbed only on the outer surfaces of the clay particles (Schnitzer, 1969; Kodama and Schnitzer, 1971). In either case metals from the clay lattice form stable complexes with the acid and are brought into solution (Schnitzer and Kodama, 1967; Kodama and Schnitzer, 1973).

As it moves down with percolating water, increasing amounts of iron and aluminum are taken up by the fulvic acid molecules. Continued removal of iron leads to the formation of a light gray albic horizon immediately below the 0 horizon. As the ratio of metal to the organic component of the organometallic complex increases, the solubility of the complex diminishes (Sokolova, 1964). At a molar ratio between iron and fulvic acid of 1:1, the complex is almost completely water soluble. As the ratio increases to 3:1, some of the complex begins to precipitate, and at a ratio of 6:1, it becomes completely insoluble (Schnitzer, 1969). Precipitation occurs entirely because of destruction of the hydrogen bond with water as a result of interaction with the metals (MacCarthy and O'Cinneide, 1974), and no special conditions of pH, flocculation with divalent cations, or oxidation–reduction relationships are needed (McKeague and St. Arnaud, 1969). Fresh fulvic acid moving through the albic horizon that has been depleted of metals reaches the previously precipitated materials, takes up some of the metal, and moves farther down the profile before it, too, precipitates. In this way, the albic horizon thickens at the expense of the spodic horizon, the spodic horizon is displaced to greater depth, and the molar ratio between metals and organic matter in the upper part of the spodic horizon is reduced relative to that in the lower part of the horizon (Fig. 16). The ratio is generally slightly higher than 3:1 in the upper part but may exceed 6:1 in the lower part (McKeague, 1968). Fulvic acid–aluminum complexes are more easily dissolved than fulvic acid–iron complexes and can migrate farther down the profile before becoming insoluble (DeKimpe and Martel, 1976). Whereas the maximum concentration of iron in a well-drained Spodosol is generally at the top of the spodic horizon together with the greatest accumulation of organic matter, the concentration of aluminum either remains constant throughout the spodic horizon or reaches its maximum in the lower part of the horizon (Brydon, 1965; DeKimpe *et al.*, 1972; DeKimpe and McKeague, 1974; Pshenichnikov, 1974; Nikonov, 1979).

In quartz sand, where little iron and aluminum is available for complexing, the albic horizon attains considerable thickness by this process rather quickly, whereas in finer materials, even where rainfall rates are very high, it may be only a few centimeters thick after long periods of development. The albic horizon is commonly variable in thickness and may have tongues extending

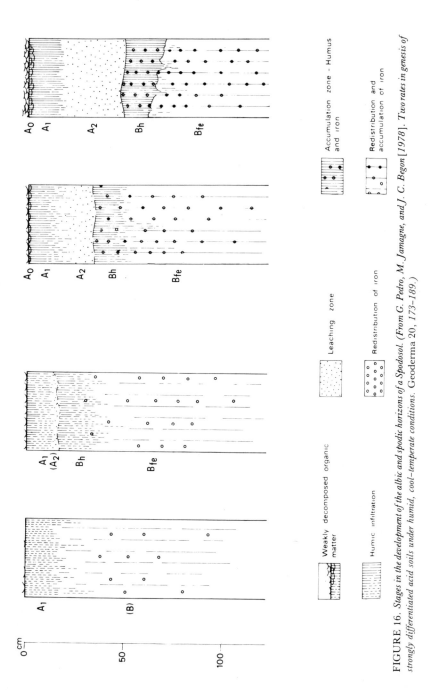

FIGURE 16. *Stages in the development of the albic and spodic horizons of a Spodosol. (From G. Pedro, M. Jamagne, and J. C. Begon [1978]. Two rates in genesis of strongly differentiated acid soils under humid, cool-temperate conditions.* Geoderma *20, 173–189.)*

into the underlying spodic horizon. It is thickest where there are concentrations of percolating water and dissolved fulvic acid because of variations in surface microrelief and in moisture pathways into the soil (Tonkonogov, 1971). As a rule, the concentration of organic matter and iron in the spodic horizon is greatest under the tongues or the thickest segments of the albic horizon.

In parent materials that are high in iron and aluminum, organometallic complexes that form are insoluble and immobile because of the wide metal to organic acid ratio, and no translocation can take place (Douchafour and Souchier, 1978; DeConinck, 1980). Soils formed in these materials are generally acidic Ochrepts.

Direct observation of soils in which the organic matter content of the upper part of the spodic horizon is very high suggest that after heavy rains organic matter in solution or colloidal suspension may pass rapidly from the O horizon to the spodic horizon through an albic horizon that had previously been depleted of its iron and aluminum. On reaching the spodic horizon, percolation slows and most of the organic matter is filtered out or precipitated (Stobbe and Wright, 1959). Highly organic coatings frequently observed on the walls of fissures in the lower part of the spodic horizon and in the underlying material in areas with very high precipitation may have a similar origin.

E. MIGRATION OF UNCOMPLEXED METALS

Although translocation in organic complexes is probably most important in the spodic process, not all of the mobile iron and aluminum in Spodosols migrates in association with fulvic acid and not all is precipitated in the spodic horizon only because of the decreasing solubility of the complexes. Some of the metals probably move in percolating water as hydrous sols (McKeague *et al.,* 1971) and some as either simple or hydroxy ions. Under reducing conditions iron can migrate as the ferrous ion, but this is unlikely in well-drained soils (McKeague, 1965). Because of the greater solubility of aluminum and hydroxy–aluminum compounds, ionic migration of aluminum exceeds that of iron (Wright and Schnitzer, 1963; Wang and Rees, 1980).

In some Spodosols the molar ratio between iron and fulvic acid in the lower part of the spodic horizon is as high as 9:1 (McKeague, 1968; McKeague and St. Arnaud, 1969). Although fulvic acid is resistant to biological degradation and tends to persist in the soil (McKeague and Sheldrick, 1977), the high ratio may be in part the result of biological destruction of the organic portion of the complex after precipitation, but ions migrating independently of organic matter may also be precipitated in the spodic horizon by biological means (McKenzie *et al.,* 1960). Bacteria, molds, fungi, and amoebas that destroy humus are common in spodic horizons, as are other microorganisms that obtain their energy by oxidizing reduced forms of iron and manganese. Iron-fixing bacteria can oxidize hundreds of times their own weight of iron and eventually are covered by thick iron deposits that may coalesce to form an ortstein layer (Aristovskaya, 1974). They are especially prevalent in Spodosols with high or fluctuating water tables.

F. MANGANESE

The 0 horizons of Spodosols are generally high in manganese because of its high concentration in the needles of coniferous trees. Manganese migrates in the soil both in combination with organic acids and in ionic form. In both forms it moves deeper than most iron and aluminum and precipitates only in the lower part of the spodic horizon, possibly because of the action of oxidizing bacteria. Where capillaries are continuous, nodules containing manganese form because of the attraction of the original precipitate for fresh manganese moving downward in solution, but in gravelly soils the manganese is evenly distributed in the lower part of the spodic horizon (Tonkonogov, 1970). In many well-drained Spodosols with high manganese concentrations in the 0 horizon, soluble manganese passes completely through the soil and does not concentrate anywhere in the solum (Moore, 1973).

G. PHOSPHORUS

In the acidic environment of Spodosols most soil phosphorus is linked to the metallic portion of the organometallic complexes (Schnitzer, 1969). Some of it probably is associated with humic acid–metal complexes in the 0 horizon and therefore is less mobile (Levesque, 1969), but a substantial proportion migrates in the soil along with soluble fulvic acid complexes. If the ratio between phosphorus and fulvic acid is high, the phosphorus and the metal may separate from the organic material as an insoluble compound (Levesque and Schnitzer, 1967). In either case the high phosphorus-fixing capacity of Spodosols reduces its availability to plants.

H. SILICA

The principal inorganic material in the soil solution of Spodosols is hydrated silica, much of it complexed with organic matter (Schnitzer and Desjardins, 1969). Its concentration is high in leachate samples from all horizons, but it passes completely through the soil and is not deposited in the spodic horizon (Singer *et al.*, 1978). The silica in solution probably originates in organic and mineral materials other than crystalline quartz, which commonly remains as a residue in horizons subject to intensive leaching.

I. CLAY MINERALS

Some clay occurs in virtually all Spodosols, although the amount may be very low in coarse sandy soils. In most cases the clay content, as measured by standard procedures, is either constant with depth or is somewhat higher in the albic horizon than in lower horizons (Targul'yan, 1964; Sokolova *et al.*, 1971), but in many Spodosols illuvial clay accumulates in the spodic horizon (Pawluk, 1960; Fedchenko, 1962; Sokolova, 1964; Brydon, 1965; Sneddon *et al.*, 1972b; Belousova *et al.*, 1973; Wang and Rees, 1980). The composition of the clay is not similar throughout the profile, as in the Alfisols, but is markedly different in the different horizons (Fig. 17). In the albic horizon the dominant clay minerals are beidellite, montmorillonite, or other expandable smectites, whereas in the spodic and lower horizons, the minerals in the clay fraction are vermiculite, chlorite, illite, amorphous hydroxides of aluminum and iron, and amorphous aluminosilicates (Yassoglov and Whiteside, 1960; McKeague, 1965; Bouma *et al.*, 1969; McKeague and St. Arnaud, 1969; Gjems, 1970; Stevens and Wilson, 1970; Sokolova *et al.*, 1971; Brydon and Shimoda, 1972; Coen and Arnold, 1972; Belousova *et al.*, 1973; DeKimpe, 1974; King and Brewster, 1976; Bain, 1977, Nicholson and Moore, 1977; Miles *et al.*, 1979; Nornberg, 1980). Kaolinite is either absent or is present only in small quantities. It may be constant throughout the profile or may occur in somewhat larger quantity in the albic horizon (Gjems, 1970; Stevens and Wilson, 1970), but even there it may be detectable only because of the absence of chlorite (DeKimpe *et al.*, 1972). It is likely that, where present, it is in large part inherited from the parent material (Belousova *et al.*, 1973; Guillet *et al.*, 1975). The normal weathering sequence in layer silicates is illite–chlorite–vermiculite–smectite–kaolinite. It is possible that the weathering process has not proceeded to the kaolinite stage because of the relative youth of most Spodosols or the low temperatures under which these soils develop (Stevens and Wilson, 1970; Sokolova *et al.*, 1971).

The occurrence of smectites in the albic horizons of sandy Spodosols formed in materials that do not originally contain those clays led Coen and Arnold (1972) to believe that they were added to the soil as atmospheric dust. It is more

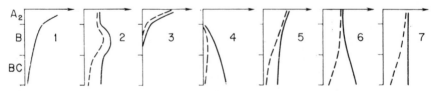

FIGURE 17. *Distribution of major components of clay fraction in Spodosols: (1) total clay; (2) amorphous iron and aluminum compounds; (3) expandable clay minerals; (4) nonswelling clay minerals; (5) mixed-layer clay minerals; (6) hydromicas; and (7) kaolinite. The solid line in segments 2–7 refers to proportions in clay fraction, and the broken line, to proportions in whole soil. (From T. A. Sokolova, V. O. Targul'yan, and G. Ya. Smirnova [1971]. Clay minerals in Fe-Al-humus podzolic soils and their role in the formation of the soil profile. Sov. Soil Sci. 3, 331–341. By permission of Scripta Publishing Co.)*

likely, however, that they are the result of intense weathering conditions immediately below the 0 horizon, where moisture and temperature variations are greatest and where the concentration of organic acids in solution is highest (Gjems, 1970; Sokolova *et al.*, 1971; Nornberg, 1980). Eluviation of the smectites to lower horizons is inhibited by flocculation resulting from the strong acidity of the soil solution (Sokolov *et al.*, 1974) and by organic acids in the interlamellar spaces and adsorbed on the outer surfaces of the clay (Yassoglou and Whiteside, 1960). These acids also act to release amorphous aluminum and iron compounds associated with the clay and to free the metals to migrate down the profile (Guillet *et al.*, 1975).

The clay fraction of the spodic horizon is high in minerals that are not as far along the weathering sequence as the smectites of the albic horizon. Less intense weathering at greater depths probably is the principal reason, but the progress of weathering may also be blocked by adsorption of organometallic complexes and amorphous iron and aluminum compounds on the clay faces (Belousova *et al.*, 1973). The spodic horizon also contains aluminosilicates in which the crystal structure is so weak or random that they appear to be amorphous under X-ray examination. The presence of substantial quantities of these clays in spodic horizons has been demonstrated by positive reaction to the Fieldes and Parrot (1966) test for allophane and allophane-like clays and by thixotropic properties that are common to all but the sandiest spodic horizons. Ferrisilicates occur in spodic horizons in lesser quantities because more of the iron is complexed with fulvic acid.

The amorphous aluminosilicates may be products of the decomposition of chlorite and vermiculite in the albic horizon that have migrated to the spodic horizon (Coen and Arnold, 1972), or they may be synthesized in the spodic horizon from aluminum and silicon hydroxides in the soil solution. They have many properties in common with the allophane derived from the weathering of volcanic ash that commonly occurs in Andepts (Wada and Harward, 1974; Chapter 8), but they differ from them in that they have lower silica to sesquioxide ratios, fewer associated water molecules, and a lower cation exchange capacity than the volcanic allophanes (Raman and Mortland, 1969).

Although the movement of fine clay in suspension is not generally an important part of the spodic process, illuviation of whole fine clay with percolating water that is largely free of organic acids may take place below the spodic horizon (Targul'yan, 1964; Belousova *et al.*, 1973). In relatively fine-grained soils, especially those developed in glacial till, this clay movement at depth probably is a major factor in the formation of a fragipan below the solum.

J. BANDING

A series of thin, reddish bands, commonly at least weakly cemented, occurs beneath the spodic horizon in many sandy Spodosols. The bands are undulating in well-drained soils but are roughly evenly spaced to depths as great as 2 m. They have no apparent relationship to groundwater levels, relief features, or bedding planes in stratified parent material. Like the spodic horizon, they are zones of concentration of iron, aluminum, and organic matter, but they may also be fairly high in exchangeable calcium and magnesium. They contain more clay than the intervening soil layers (Wurman *et al.*, 1959; Nikitin and Fedorov, 1977). In acidic parent materials the bands probably originate in much the same way as spodic horizons, in that soluble fulvic acid that escaped precipitation in the spodic horizon combines with iron and aluminum as it percolates downward and precipitates in part as the ratio of metal to organic matter increases. The process then repeats with depth until all of the fulvic acid is exhausted. In originally basic material, the site of precipitation may have been determined by the upper boundary of a layer with a high concentration of bases; in time, after most of the bases had been leached to greater depth and the fulvic acid was again soluble and active, complexing and downward migration resumed until the precipitation ratio was attained or until the acid was again immobilized by a high base concentration. The clay in the bands may have migrated in association with the organometallic complexes and precipitated with them, or may have moved independently in suspension until it was flocculated on coming into contact with the organometallic band (Folks and Riecken, 1956). Bands high in iron that are formed in Spodosols with high or fluctuating water tables are discussed under Cryaquods later in this chapter.

K. BRITTLENESS

Spodic horizons, especially those in sandy soils, become increasingly brittle in time. The degree of brittleness is related in part to the kind of vegetation growing on the soil. In Swiss alpine areas, for example, sandy soils under native heath have brittle spodic horizons, but where the soils have been maintained in pasture grasses the spodic horizons are friable (Buoma *et al.*, 1969). The brittleness is attributed to connected coatings of precipitated organometallic complexes on sand grains (McKeague and Wang, 1980). This condition can develop only where there is insufficient biological activity, such as root growth or animal burrowing, to mix illuviated organometallic complexes with the mineral matrix (DeConinck, 1980). The coatings may become thick enough to form a cemented layer, or ortstein, or may crumble to silt-size pellets. Brittleness may also be related to extreme weathering of primary minerals such as muscovite in the upper horizons of the soils and the subsequent formation of clay bridges in the spodic horizon. Other studies suggest that aluminum oxides may be responsible for aggregation in spodic horizons (Saini *et al.*, 1966).

II. Identification of Spodic Horizons

A. MORPHOLOGICAL CRITERIA

In general, spodic horizons and, therefore, Spodosols can be identified readily in the field. The principal field criteria are black to dark reddish brown colors in the upper part of the horizon, usually below an abrupt wavy or irregular boundary with an albic horizon, grading to more yellowish colors in the lower part of the horizon; weak structure except for fine granules; and, in the case of sandy soils, organometallic pellets or coatings on sand grains. Clay skins on peds or in pores are absent. There are situations, however, in which more than field criteria are needed for positive recognition of a spodic horizon. In some soils, especially in alpine areas, the albic horizon that normally overlies the spodic horizon may be obscured by organic matter. In cultivated soils the albic horizon may have been thoroughly mixed with the upper part of the spodic horizon, leaving only the lower part of the horizon unaltered. Cambic horizons in ashy soils that have properties much like those of spodic horizons may be difficult to distinguish from horizons in the same material that have formed as a result of the spodic process. Other soils have albic horizons over brown cambic horizons with little or no accumulation of illuvial materials but with a superficial resemblance to weakly developed spodic horizons. In soils in which there has been some illuviation of fine clay, specific criteria are needed to define the limits of the spodic horizon with respect to the argillic horizon. Finally, criteria are needed to distinguish spodic horizons in poorly drained soils from strongly mottled cambic horizons that are high in iron.

B. CHEMICAL CRITERIA

Chemical criteria have been developed for identification of the spodic horizon in such soils but, in every case, these criteria are used only when positive identification is not possible from morphological criteria alone.

The several forms of iron and aluminum in soils can be identified by the use of different extracting solutions (Fig. 18). A dilute solution of sodium pyrophosphate at pH 10 extracts only the iron and aluminum complexed with organic acids and relatively small amounts of the metals in amorphous hydrous oxides or that are adsorbed on clay surfaces. Acid ammonium oxalate extracts most of the pyrophosphate-extractable material and, in addition, iron and aluminum in weakly crystallized hydrous oxides and in easily weatherable primary minerals. Dithionite-citrate extracts all of these plus the iron and aluminum in well-crystallized oxides (McKeague, 1967; Bascomb, 1968; Lutwick and Dormaar, 1973). In all cases the extractants are more specific for iron than for aluminum (McKeague *et al.,* 1971; Pawluk, 1972).

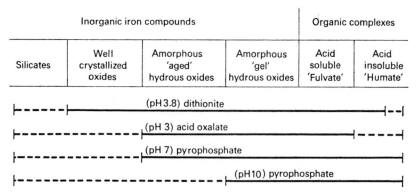

FIGURE 18. *Forms of iron removable from soils by different extractants. (From Bascomb [1968] by permission of British Society of Soil Science.)*

In the United States Taxonomy, chemical criteria for identification of spodic horizons that cannot be positively identified visually are designed to reflect the fact that the essential feature of the spodic process is the illuviation of iron and aluminum in complex association with organic acids. The absolute quantity of pyrophosphate-extractable iron and aluminum is not used as a sole criterion because, in loamy soils with high organic matter content, some of the extracted metals may not have migrated as part of an organometallic complex. The ratio between pyrophosphate-extractable iron and aluminum and the measured clay percentage,* however, should be narrower in a spodic horizon than in unaltered soil material or in argillic or cambic horizons (Franzmeier *et al.*, 1965; Blume and Schwertmann, 1969; DeKimpe, 1970; Moore, 1973; King and Brewster, 1976). For confirmation of the existence of a spodic hor- izon where morphological criteria are not definitive, this ratio currently is re- quired to be at least .2,† but there are indications that a wider ratio may be more desirable. Loamy soils with morphological and, in other respects, chemical properties of Spodosols in Quebec (DeKimpe and Martel, 1976) and western Washington (Singer *et al.*, 1978), have ratios between .1 and .2. Similar sandy soils in New Brunswick have ratios ranging from .05 to .18 (Wang and Rees, 1980). It seems likely that soils in which profile characteristics are less clearly those of Spodosols may also have ratios in that range. In the Canadian soil classification system, the required ratio for Podzolic B horizons (equivalent to the spodic horizon), except for those in which the illuvial material is almost

---

* *Clay percentage* refers to clay determined by the standard pipette procedure of the United States Soil Survey Laboratory. Because soils containing amorphous mineral material may not disperse completely, this value may be lower than the actual proportion of clay-size particles in the soil.

† Where pyrophosphate-extractable iron is less than .1%, as in some poorly drained soils, the ratio of pyrophosphate-extractable carbon plus aluminum to the percentage of clay is substituted; this ratio is also required to be at least .2.

completely organic, is only .05 (Canada Soil Survey Committee, 1978). In Scotland, on the other hand, most soils with the gross morphological characteristics of Spodosols have ratios narrower than .2, but a substantial number of soils that do not exhibit those characteristics also meet this criterion, primarily because they are very low in clay (Ragg *et al.*, 1978). The existence of a spodic horizon in most of these soils, however, is indicated by the presence of amorphous organometallic pellets and coatings on sand grains.

It is also required for identification of a spodic horizon where other criteria are not conclusive that the iron and aluminum extractable by sodium pyrophosphate be at least one-half of that extractable by dithionite-citrate—that is, that at least 50% of the extractable iron and aluminum in the horizon is derived from organometallic complexes or amorphous hydrous oxides. This test is intended to distinguish between spodic horizons and cambic horizons in ashy materials that have many of the same physical and chemical properties but in which illuviation of organometallic complexes has not been significant. Unfortunately, in cold areas this test may not always result in a clear separation of the two kinds of soil. In soils formed in fine ash on Kodiak Island, Alaska, for example, the ratio between the two extracts is considerably narrower than .5, although the soils do not have the morphological characteristics of Spodosols. These soils are very high in organic matter, however, and this may have masked morphological features of Spodosols that might otherwise have developed. Elsewhere in Alaska, however, albic horizons and spodic horizons that satisfy both the morphological and chemical criteria form readily in ashy materials (Rieger, 1974). In other soils this criterion may be too restrictive. In 21 spodic horizons of Canadian soils, for example, pyrophosphate extracted on average less than half as much iron and aluminum as dithionite (McKeague *et al.*, 1971) (Fig. 19). Ragg *et al.* (1978) believe that a ratio of .36 would most effectively differentiate spodic from nonspodic soils in Scotland.

The high pH-dependent cation exchange capacity that is characteristic of organometallic complexes (Clark *et al.*, 1966) is used to establish the minimum degree of expression required for recognition of a spodic horizon.* The cation exchange capacity determined at pH 8.2, adjusted both for the portion of the exchange capacity that is associated with clay and for the thickness of the horizon, must exceed a minimum "index of accumulation" value. This procedure makes it unnecessary to require minimum concentrations of iron, aluminum, and organic carbon or a minimum thickness for the horizon. Many cambic and argillic horizons also exceed the threshold index of accumulation, but they do not satisfy the other criteria for the spodic horizon. It has been sug-

---

* The pH-dependent cation exchange capacity is the difference between the cation exchange capacity measured by a neutral extracting solution, such as ammonium acetate buffered at pH 7, and that measured by an extracting solution at higher pH, usually 8.2.

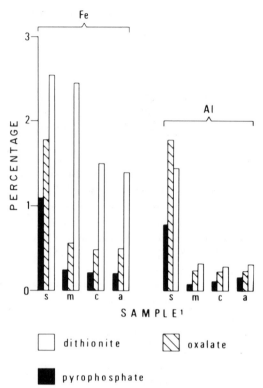

FIGURE 19. *Mean amounts of iron and aluminum extracted from various kinds of horizons in Canadian soils by dithionite, oxalate, and pyrophosphate: (s) 21 spodic horizons; (m) 13 mottled B horizons from Aqualfs and Aquepts; (c) 9 brown cambic horizons from well- or moderately well-drained soils; and (a) 7 argillic horizons from Boralfs and Borolls (McKeague* et al., *1971). (Reprinted from* Soil Science Society of America Proceedings, *Volume 35, page 35, 1971. By permission of the Soil Science Society of America.)*

gested that measurement of the pH-dependent cation exchange capacity that can be attributed to organic matter alone may be adequate by itself to separate spodic horizons from other diagnostic horizons in strongly acidic soils (Clark and Nichol, 1968).

Although not presently recognized in any classification system, it is possible that relationships between humic acid, fulvic acid, and the phenolic acid component of fulvic acid could also be used in the identification of spodic horizons. Phenolic acid generally makes up more than 40% of the fulvic acid in spodic horizons and much less than 40% in the fulvic acid of argillic horizons. The fulvic acid of brown cambic horizons may also be high in phenolic acid, but the proportion of humic acid is generally higher than in spodic horizons, where the ratio between humic acid and fulvic acid seldom is narrower than 1:3 (Lowe, 1980).

III. Classification of Cold Spodosols

A. ORTHODS

The ratio between iron extractable by dithionite-citrate and organic carbon in the upper part of the spodic horizon is used in the United States Taxonomy to separate well-drained Spodosols in which the dominant illuvial material is either iron or organic matter from those in which neither clearly dominates. Aluminum translocation occurs in all Spodosols, but apparently there is no situation in which it accumulates in a spodic horizon without simultaneous accumulation of either iron or organic matter, or both. By far the greatest number of well-drained Spodosols have spodic horizons in which organically complexed iron is a prominent, if not the principal, accumulation product. These soils are identified by a ratio between dithionite-extractable iron and organic carbon between 1:5 and 6:1. All of them except soils underlain by fragipans or that contain placic horizons within the spodic horizon are classified in cold regions as Cryorthods; those with fragipans are separated in a subgroup of Cryic Fragiorthods, and those with placic horizons in an as yet undifferentiated group of Placorthods.

Typic Cryorthods have either 1.2–6% organic carbon (approximately 2–10% organic matter) in the upper part of the spodic horizon or a cemented or indurated spodic horizon. These are the most extensive Cryorthods. At low elevations in cold regions, they occur mostly under coniferous or mixed forests. They have no permafrost but may occur in geographic association with poorly drained soils with permafrost. No precise range in annual precipitation rates can be given because the existence of a leaching moisture regime depends on the nature of the parent material and the air and soil temperature regime, as well as on the quantity of rainfall. In loamy, nonashy materials, however, Typic Cryorthods generally give way to Cryochrepts in strongly continental climates where annual precipitation is less than 350–400 mm and to Humic Cryorthods in maritime climates where annual precipitation is greater than 1500 mm. In coarse-grained materials that are low in bases, they can develop where annual precipitation is as low as 200 mm. In mountainous areas with no permafrost, Typic Cryorthods form under forest on lower slopes and under tundra or grassy vegetation above tree line. In all cases the intensity of development in similar parent materials increases with increasing precipitation (Rieger and DeMent, 1965; Pshenichnikov, 1974). Their occurrence is also affected by evaporation rates. In mountains of southeastern Siberia, for example, they occupy southern slopes in the cooler northern sections but exist only on north-facing slopes in the southern sections where regional evaporation rates are higher (Ufimtseva, 1963).

The soils form readily in materials like quartz sand or weathered granite, but their development is retarded in materials derived from basic rocks. In the Siberian mountains soils with Spodosol morphology occur in granitic material, but soils formed in material derived from amphibolite under identical climate conditions and vegetation do not have those characteristics. In part this is a result of the dark color of the amphibolite, which masks the albic and spodic horizons, but the degree of translocation of iron and organic matter is also lower in the more basic soils (Sokolova, 1964). Cryorthods formed in volcanic materials have more distinct horizonation and a higher proportion of amorphous minerals in the spodic horizon (Naumov *et al.*, 1964).

The solum of Typic Cryorthods is generally thin. It seldom extends deeper than 50 cm even in coarse materials and may be less than 30 cm thick. The albic and spodic horizons are always strongly to extremely acid. Minimum pH values occur, generally, at the bottom of the albic horizon (Pawluk, 1960; Mikhaylov, 1970; Ugolini *et al.*, 1977), but because of its high cation exchange capacity, minimum base saturation (which is very low throughout the solum) is in the spodic horizon (Pawluk, 1960; Sneddon *et al.*, 1972b). Carbon to nitrogen ratios are very wide in the 0 horizon and normally exceed 15 in the solum (Rieger and DeMent, 1965). Under grass–forb vegetation in alpine areas, however, the carbon to nitrogen ratio may be fairly narrow (Sneddon *et al.*, 1972b). Available phosphorus below the 0 horizon is low.

Humic Cryorthods have more than 6% organic carbon (10% organic matter) in the upper 10 cm of the spodic horizon. They occur, principally, under maritime climates with very high precipitation and long frost-free periods. In southeastern Alaska, eastern Canada, and coastal areas of eastern Siberia, they are forested. In other areas, such as northern Scotland, they develop under vegetation dominated by heath, lichens, and sedges at high elevations and under a blanket of moss peat in a lower zone. In acidic materials the upper part of the spodic horizon in Scotland contains about 10% organic carbon, but in basic igneous material it is as high as 24% (Romans *et al.*, 1966). Organic matter in the spodic horizon of Humic Cryorthods coats stones and, in the lower part of the horizon, commonly occurs as discontinuous lenses or as linings on the walls of fissures (Heilman and Gass, 1974).

These soils develop under dense vegetation and conditions that favor high biological activity and intensive leaching. As a result, the characteristics of Typic Cryorthods are intensified. The soils are extremely acid, have very high cation exchange capacity and very low base saturation in the spodic horizon even in soils formed from originally calcareous material, and are strongly thixotropic. Many have thick 0 horizons in which the lower part is highly decomposed. Plants growing on them have shallow root systems, and recycling of nutrients is largely confined to the 0 horizon (Heilman and Gass, 1974). In many of these soils, the high production of colored organic acids tends to obscure the albic horizon so that, despite the strong development of the spodic horizon, it is very thin or may not be readily visible (Pshenichnikov, 1976). In some Humic Cryorthods the proportion of organic acids that is complexed with aluminum exceeds that complexed with iron (Wicklund and Whiteside, 1959; Pshenichnikov, 1974). In this respect, as well as in the amount of organic matter precipitated in the spodic horizon, the soils are closely related to the great group of Cryohumods.

Boralfic Cryorthods are bisequal, with a thin albic–spodic horizon sequence in the upper part of the soil and an argillic horizon at greater depth. Usually a second albic horizon separates the spodic horizon from the underlying argillic horizon. These soils are widespread in eastern Canada in glacial till derived from calcareous shale and sandstone (McKeague *et al.,* 1967a; McKeague and Cann, 1969; McKeague *et al.,* 1973) and also occur in some mountainous areas elsewhere. The initial process in their development was leaching of carbonates followed by translocation of whole fine clay to form an argillic horizon beneath a thick albic horizon. Depletion of clay and the removal of bases from the albic horizon created conditions favorable for initiation of the spodic process and, in time, the formation of a spodic horizon in the earlier albic horizon (McKeague and Cann, 1969). In some cases translocation of organic matter and iron into the spodic horizon and clay translocation into the lower argillic horizon apparently are taking place simultaneously (Cann and Whiteside, 1955; Pedro *et al.,* 1978); that is, the spodic process has started before cessation of the initial process of clay illuviation.

Pergelic Cryorthods are common in high mountains and at high latitudes under both coniferous forest and heath vegetation and occur in a few places in the arctic tundra. They have mean annual temperatures of 0° C or less, but the permafrost table is always deep enough so that any water perched above the permafrost has no effect on the spodic process. Depth to ice-rich permafrost ranges from about 1 m to more than 6 m (Filimonova, 1965; Nicholson and Moore, 1977), but dry permafrost may occur at shallower depths. The solum in these soils is seldom thicker than 25 cm and may be as thin as 10 cm. Thickness and intensity of development of the solum is related directly to temperature and precipitation rates (Karavayeva *et al.,* 1965). Where the soil surface is hummocky, there is greater accumulation of iron and organic matter in the spodic horizon in the troughs between hummocks than at the tops of the hummocks (Koposov, 1976).

In many mountainous areas Pergelic Cryorthods are the dominant soils, but they are almost always associated with Cryaquepts over shallow permafrost in depressions and on steep polar-facing slopes. With increasing elevation and colder temperatures, biological and chemical processes slow to the point that illuviation of organometallic compounds does not take place, and the Cryorthods give way to Pergelic Cryumbrepts or Cryochrepts (Kubota and Whittig, 1960). In some areas with variable soil textures, Cryorthods form in coarse materials, and Cryumbrepts or Cryochrepts form in finer materials (Karavayeva *et al.*, 1965). In the arctic tundra Cryorthods form only in coarse-grained materials (Kreida, 1958; Ugolini, 1966) or in shallow depressions or sheltered slopes where snow accumulates—that is, where soil temperatures are somewhat higher and there is adequate moisture for leaching (Brown and Tedrow, 1964; Brown, 1966).

Many Pergelic Cryorthods are subject to severe cryoturbation. This may result in undulating and broken soil horizons (Filimonova, 1965; Nicholson and Moore, 1977) or in the development of barren frost circles. Where barren areas are extensive and the spodic horizon that formed under more stable conditions has been destroyed or broken into discontinuous segments, the soils are classified as *Pergelic Ruptic-Entic Cryorthods.* In areas where precipitation rates are barely high enough for Spodosols to form, however, there is little frost action (Kuz'min and Sazonov, 1965).

In southeastern Alaska Pergelic Cryorthods have developed in vegetated debris that covers stagnant ice at the mouths of retreating glaciers (Stephens, 1969). These soils eventually collapse as the ice melts, but the products formed as a result of the spodic process facilitate the development of Spodosols in the newly formed moraines.

Other subgroups of Cryorthods are less extensive. *Entic Cryorthods,* which have less than 1.2% organic carbon (2% organic matter) in the spodic horizon, occur mostly in freshly exposed or deposited materials such as sand dunes that have recently been stabilized by vegetation. In time, virtually all will become Typic or Pergelic Cryorthods. *Lithic Cryorthods* overlie consolidated bedrock less than 50 cm below the surface of the mineral soil and may or may not have a pergelic temperature regime. They occur principally on steep mountain slopes and high ridges and support either forest or alpine tundra vegetation. *Humic Lithic Cryorthods* are similar to Humic Cryorthods except for shallow bedrock. They commonly occur in association with Cryofolists in areas with very high precipitation (Rieger *et al.*, 1979).

Fragiorthods have a fragipan at some depth below the spodic horizon. All Orthods with fragipans in cold regions, except those with a placic horizon above or within the spodic horizon, are classified as *Cryic Fragiorthods.* A placic horizon may occur below the spodic horizon but above the fragipan. The solum may have characteristics of either Typic Cryorthods or Humic Cryorthods. The soils form mostly in loamy glacial till in areas with high rainfall (Wang *et al.*, 1974). The fragipans are rarely calcareous but commonly are less acidic than the overlying solum. In areas with calcareous glacial till, they form only after leaching of the carbonates and acidification of the till. Except in sandy materials there is generally an albic horizon between the bottom of the spodic horizon and the surface of the fragipan, caused by lateral moisture flow over the slowly permeable pan (Mehuys and DeKimpe, 1976) (Fig. 20).

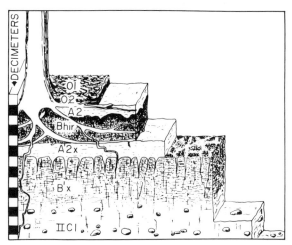

FIGURE 20. *Sequence of horizons in a Fragiorthod from Wisconsin. In soils of colder climates, it is unlikely that any tree roots would penetrate to the fragipan. (From Grossman and Carlisle [1969].)*

As in most fragipans, the firmness of the pan in Fragiorthods probably is the result primarily of illuviation of fine, nonexpansible clay from upper horizons and the formation of clay bridges between coarser particles (Grossman and Carlisle, 1969). There are indications, however, that under Spodosols it may be due at least in part to products of the spodic process and that development of the fragipan goes on simultaneously with development of the spodic horizon (Yassoglou and Whiteside, 1960). For example, investigations in materials that can be dated by their relationship to construction of an earthen bank during the Roman occupation of Britain show that the indurated horizon started to form soon after the beginning of development of a spodic horizon in originally calcareous till (Romans, 1962). Precipitation of alumina and, to a lesser extent, iron oxides from the soil solution apparently adds to the firmness of the pan (Yassoglou and Whiteside, 1960). Very little of the iron and aluminum in the fragipans is associated with organic matter. Much of the iron is dithionite-extractable, and the ratio between pyrophosphate-extractable iron and clay is very wide (DeKimpe, 1970; Wang *et al.*, 1974). At least some fragipans under spodic horizons are high in oxalate-extractable aluminum (DeKimpe, 1970). Coatings on clay particles of amorphous aluminum, iron, and silicon oxides and hydroxides have been identified as major bonding agents in fragipans of nonspodic acid soils (Hallmark and Smeck, 1979).

In sandy materials and in soils formed at least in part in volcanic materials, a dense, compacted pan that resembles, but is not, a fragipan may occur under Spodosols. The cementing material is believed to consist of amorphous to weakly crystalline aluminosilicates or ferrosilicates that were precipitated from solution because of oxidizing conditions or higher pH below the spodic horizon (McKeague and Sprout, 1975; Moore, 1976). No provision is made in the United States Taxonomy at present for separate classification of soils with this kind of pan.

Placorthods have an undulating placic horizon within the spodic horizon. They occur only in areas with high precipitation rates and climates that permit effective leaching during most of the year. In the cryic zone they are known to occur in Newfoundland (McKeague *et al.*, 1967b; McKeague *et al.*, 1968) and in the northern British Isles (Proudfoot, 1958; Ragg and Ball, 1964). They have been observed in coastal British Columbia where temperatures are somewhat warmer than cryic (Lavkulich *et al.*, 1971), but in southeastern Alaska placic horizons apparently exist only in soils derived from volcanic ash that are classified as Humods and in poorly drained Aquods. All Placorthods have the high organic matter content in the spodic horizon that is characteristic of Humic Cryorthods.

The placic horizon in these soils, although usually less than 1 cm thick, normally consists of two distinct layers—a black upper layer dominated by fulvic acid complexed with iron and a reddish lower layer that is high in crystalline iron oxide. It contains a much higher proportion of iron relative to aluminum than adjacent portions of the spodic horizon (McKeague *et al.*, 1967b). It has been suggested that the excess iron in the placic horizon has migrated in the ferrous form from the soil above the horizon as a result of saturated conditions induced by the presence of the impermeable placic horizon itself (Flach *et al.*, 1969), but this does not account for the original formation of the horizon. Although the processes involved in the development of placic horizons are not fully understood, it is noteworthy that most of them in cold regions are associated with a textural discontinuity in the parent material.

B. HUMODS

Cryohumods are cold, well-drained Spodosols in which, at least in part of the spodic horizon, the ratio between dithionite-extractable iron and organic carbon is wider than 1:5 and that have no placic horizon in the spodic horizon and no fragipan below the solum. *Typic Cryohumods* have spodic horizons that are high in organic matter, but in the absolute quantity of organic matter and in their appearance they are indistinguishable from Humic Cryorthods (Rieger, 1974). They occur in areas with moderate to high rainfall under coniferous forest, heath, and, in some cold maritime climates, grassy vegetation.

Cryohumods appear to be restricted to parent materials that are deficient in iron, such as quartz sand and some volcanic ash deposits. In quartz sand, where almost no iron or aluminum other than that in the organic litter is available for complexing, it seems likely that much of the organic matter in the spodic horizon has migrated from the 0 horizon as a colloidal suspension (Stobbe and Wright, 1959). In volcanic ash, which is low in iron but in which aluminum is readily available as a weathering product, aluminum is the principal complexing metal. Cryohumods in ashy material also are very high in amorphous aluminosilicates, much of which probably are derived directly from weathering of the ash.

Cryohumods in ashy materials are not always easily identified in the field because, where the content of organic material is very high, the albic horizon may be obscured. Even a thin albic horizon below the 0 horizon can be considered to be a clear indication of the operation of the spodic process. Where the albic horizon is absent, only the ratio between pyrophosphate- and dithionite-extractable iron and aluminum can be used to determine whether or not a spodic horizon exists and whether the soil is properly classified as a Humod or an Andept. As suggested earlier, it is not yet certain that this criterion is valid.

Lithic Cryohumods, with bedrock within 50 cm of the mineral surface, are inextensive but have been observed on high ridges above tree line in southeastern Alaska, in areas dominated by organic soils. Although provision is made for a subgroup of *Pergelic Cryohumods,* with mean annual soil temperatures of 0° C or less, in the United States Taxonomy, no such soils have yet been described. Humods with a placic horizon in the spodic horizon have formed in old volcanic ash deposits in southeastern Alaska (Flach *et al.,* 1969); these soils are classified as *Cryic Placohumods.* No Humods with fragipans have yet been identified in cold regions. If they exist they would be classified as *Cryic Fragihumods.*

C. FERRODS

Ferrods have thick albic horizons and reddish spodic horizons that are high in iron and low in organic matter (the ratio between dithionite-extractable iron and organic carbon is wider than 6:1) and that commonly are strongly cemented. They are believed to occur only in cold continental areas with low precipitation, and there only in sandy or gravelly materials near the margins of depressions containing intermittent lakes that dry out by midsummer. Adjacent upland soils may be Typic Cryorthods in sandy materials but are Cryochrepts in loamy materials. Although the Ferrods are forested, it is likely that slow decomposition rates of the forest litter under dry summer conditions results in a low content of soluble organic acids. Precipitation of iron in the spodic horizon probably is directly from solution as the lake level and the water table in the soils recede during the summer. In similar situations where the parent materials are high in soluble salts, saline soils rather than Ferrods surround the depressions. Because these soils are very inextensive and have not been studied, no classification below the suborder level has been developed.

D. AQUODS

Cryaquods are Spodosols of cold areas that are saturated long enough during the summer so that dissolved oxygen is depleted. They exhibit characteristics of wetness such as a histic epipedon, mottling in the upper part of the profile, and iron–manganese concretions below the spodic horizon. Aquods with placic horizons or fragipans are excluded from the great group of Cryaquods. The vegetation ranges from a forest of conifers and birch similar to that on adjacent well-drained Spodosols but with a more peaty or grassy forest floor to shrub–lichen or grassy tundra. The soils commonly occupy slopes subject to seepage and intermediate positions between well-drained uplands and bogs. In some mountainous areas with high rainfall, they form over shallow bedrock that prevents percolation of excess moisture (Bouma and Van der Plas, 1971). In the plains of western Siberia, they are the dominant soils of low interfluves between areas dominated by Cryaquepts (Karavayeva, 1973). They may occur in loamy materials where well-drained Spodosols form in sandy, more permeable materials under identical climatic conditions (Koposov, 1976) or in sandy deposits underlain by less permeable materials (Rusanova, 1979).

Typic Cryaquods consist of a relatively thick albic horizon over a spodic horizon that is high in organic matter and that commonly is made up of a series of indurated horizontal slabs. They form mostly in sandy materials at the margins of strongly acidic lakes and bogs or downslope from upland peaty deposits in humid climates. The iron and manganese content of both the albic and spodic horizons is low. Under saturated conditions organic matter moves downward, probably in complex association with aluminum (Miles *et al.*, 1979) and precipitates during periods of recession of the water table to form a humic spodic horizon. Iron is converted to the ferrous form and, along with soluble forms of manganese, is leached to considerable depth (Damman, 1962). A strongly cemented horizon high in iron may occur below the humic spodic horizon if oxidizing conditions exist at that depth when water levels are low, but much of the iron is desseminated throughout the bog and eventually deposited at its base as a result of the activity of oxidizing microorganisms. In Cryaquods with indurated spodic horizons, the lower part of the albic horizon may be mottled because of precipitation of organometallic complexes from perched water (Nikonov, 1979).

Although soils with humic spodic horizons are designated the Typic subgroup of Cryaquods, they are not the most extensive Cryaquods of cold regions. Probably most extensive are the *Sideric Cryaquods,* which occupy shallow depressions in uplands, slopes subject to seepage, and level or gently sloping uplands in areas with fine-grained parent materials. These soils are saturated and subject to gleying during the early part of the summer but are more or less freely drained in late summer, when it is possible for at least a modified form of the spodic process to operate (Nicholson and Moore, 1977).

Thawing is less rapid than in well-drained Spodosols because of the large quantities of available water and the thicker, more peaty organic mat on the surface of the soils. Moisture perched above the slowly thawing subsurface layer creates reducing conditions in the upper part of the profile, and even after thawing is complete, the upper horizons remain at or near field capacity for a considerable time. This leads to mottling in loamy soils where saturated and unsaturated pockets exist simultaneously (McKeague, 1965). Iron- and manganese-fixing bacteria, which are especially numerous in wet soils and which grow best at the interface between the reduced soil solution and atmospheric oxygen in the unsaturated pockets and larger pores, probably play a major role in this process. In sandy soils with a shallow oxygen-depleted water table that remains at a constant depth for long periods, a band of high iron concentration forms at the contact zone (Aristovskaya, 1963). The band may be strongly cemented, especially where periodically aerated, coarser-textured material overlies a finer perennially reduced subsoil (Evans *et al.*, 1978). A sequence of horizontal bands may develop if the water table fluctuates periodically and the soil is exposed alternately to reducing and oxidizing conditions (Tonkonogov, 1970). In some cases water that percolates into a recently thawed subsoil encounters an oxidizing environment because of air that had been trapped during freezing, and the ferrous iron in solution is rapidly precipitated (Bouma and Van der Plas, 1971; Ignatenko, 1972). Similarly, under conditions where water beneath the water table originated upslope and is aerated, reducing conditions may prevail in the upper part of the profile and oxidizing conditions below the water table. Ferrous iron reaching the water table is then oxidized and precipitated (McKeague, 1965). Aluminum has greater mobility than iron in these situations and accumulates at greater depth (Kreida, 1958).

Eluviation and accumulation of iron is the major genetic characteristic of Sideric Cryaquods (Karavayeva, 1973). Organic matter, although it participates in the illuvial process, accumulates only slightly in the upper part of the spodic horizon. It generally is more evenly distributed in the soil than in the well-drained Spodosols and decreases gradually with depth (Rozhnova and Schastnaia, 1959; Rubtsov, 1964; Karavayeva, 1973; Koposov, 1976). In areas with severe winters, much of it is dispersed through the soil by cryoturbation (Kuz'min and Sazonov, 1965; Zaboyeva, 1965).

Pergelic Sideric Cryaquods have permafrost at some depth, but the permafrost table in these soils is always deep enough so that the soils are not waterlogged throughout the summer. Some of them are forested, but most support low shrub or tundra vegetation. Associated soils with shallow permafrost, usually in finer materials or in lower positions, do not develop spodic horizons.

Aquods with fragipans but with no placic horizon in cold regions are classified as *Cryic Fragiaquods*. These soils are not extensive because less intense clay weathering and smaller amounts of clay illuviation below the spodic horizon in poorly drained soils than in soils with deeper water tables make it less likely that fragipans will form (DeKimpe, 1976). They do exist, however, as a substratum of Aquods developed in loamy or clayey glacial till in areas with very high precipitation.

Aquods with a placic horizon above the spodic horizon or above a fragipan whether or not there is an overlying spodic horizon are *Cryic Placaquods*. They generally occur in poorly drained depressions in areas with strongly maritime climates. In soils in which material of somewhat finer texture overlies a coarser substratum, the placic horizon forms at the contact between the two parent materials, possibly because of precipitation of iron and organic matter at the interface between reducing conditions in the finer material and oxidizing conditions in the substratum (Valentine, 1969). If the placic horizon occurs in or below the spodic horizon in cold Aquods without fragipans, the soils are classified as *Placic Haplaquods* despite their cryic temperature regime. This quirk in the classification system probably will be corrected in future revisions.

References

Alekseyeva, N. S., and Pereverzev, V. N. (1974). Effect of cultivation on the composition of organic matter in the illuvial humic Podzols of the Kola Peninsula. *Sov. Soil Sci.* **6**, 471–480.

Anderson, H. A., Fraser, A. R., Hepburn, A., and Russell, J. D. (1977). Chemical and infrared spectroscopic studies of fulvic acid fractions from a Podzol. *J. Soil Sci.* **28**, 623–633.

Archegova, I. B. (1968). Sandy soils of the tundra subzone (Vorkuta region). *Sov. Soil Sci.,* 902–904.

Aristovskaya, T. V. (1963). Decomposition of organic mineral compounds in podzolic soils. *Sov. Soil Sci.,* 20–29.

Aristovskaya, T. V. (1974). Role of microorganisms in iron mobilization and stabilization in soils. *Geoderma* **12**, 145–150.

Bain, D. C. (1977). The weathering of ferruginous chlorite in a Podzol from Argyllshire. *Geoderma* **17**, 193–208.

Bascomb, C. L. (1968). Distribution of pyrophosphate-extractable iron and organic carbon in soils of various groups. *J. Soil Sci.* **19**, 251–268.

Bel'chikova, N. P. (1966). Characteristics of humic substances of taiga soils in central Siberia which have developed on basic rocks. *Sov. Soil Sci.,* 1136–1145.

Belousova, N. I., Sokolova, T. A., and Tyapkina, N. A. (1973). Profile differentiation of clay minerals in Al–Fe–podzolic soils on granite. *Sov. Soil Sci.* **5**, 692–708.

Bloomfield, C., Kelso, W. I., and Pruden, C. (1976). Reactions between metals and humified organic matter. *J. Soil Sci.* **27**, 16–31.

Blume, H. P., and Schwertmann, U. (1969). Genetic evaluation of profile distribution of aluminum, iron, and manganese oxides. *Soil Sci. Soc. Am. Proc.* **33**, 438–444.

Bouma, J., and Van der Plas, L. (1971). Genesis and morphology of some alpine pseudogley profiles. *J. Soil Sci.* **22**, 81–93.

Bouma, J., Hoeks, J., Van der Plas, L., and Van Scherrenburg, B. (1969). Genesis and morphology of some alpine Podzol profiles. *J. Soil Sci.* **20**, 384–398.

Brown, J. (1966). "Soils of the Okpilak River Region, Alaska." *U. S. Army Cold Reg. Res. Eng. Lab. (CRREL) Res. Rep. 188.* Hanover, New Hampshire.

Brown, J., and Tedrow, J. C. F. (1964). Soils of the northern Brooks Range, Alaska: 4. Well-drained soils of the glaciated valleys. *Soil Sci.* **97**, 187–195.

Brydon, J. E. (1965). Clay illuviation in some Orthic Podzols of eastern Canada. *Can. J. Soil Sci.* **45**, 127–138.

Brydon, J. E., and Shimoda, S. (1972). Allophane and other amorphous constituents in a Podzol from Nova Scotia. *Can. J. Soil Sci.* **52**, 465–475.

Buurman, P., Van der Plas, L., and Slager, S. (1976). A toposequence of alpine soils on calcareous micaschists, northern Andula region, Switzerland. *J. Soil Sci.* **27**, 395–410.

Canada Soil Survey Committee, Subcommittee on Soil Classification. (1978). "The Canadian System of Soil Classification." Can. Dept. Agr. Pub. 1646. Supply and Services Canada, Ottawa.

Cann, D. B., and Whiteside, E. P. (1955). A study of the genesis of a Podzol-Gray-Brown Podzolic intergrade soil profile in Michigan. *Soil Sci. Soc. Am. Proc.* **19**, 497–501.

Chen, Y., and Schnitzer, M. (1976). Scanning electron microscopy of a humic acid and of a fulvic acid and its metal and clay complexes. *Soil Sci. Soc. Am. Jour.* **40**, 682–686.

Chen, Y., Khan, S. U., and Schnitzer, M. (1978). Ultraviolet irradiation of dilute fulvic acid solutions. *Soil Sci. Soc. Am. Jour.* **42**, 292–296.

Clark, J. S., and Nichol, W. E. (1968). Estimation of the inorganic and organic pH-dependent cation exchange capacity of the B horizons of Podzolic and Brunisolic soils. *Can. J. Soil Sci.* **48**, 53–63.

Clark, J. S., McKeague, J. A., and Nichol, W. E. (1966). The use of pH-dependent cation-exchange capacity for characterizing the B horizons of Brunisolic and Podzolic soils. *Can J. Soil Sci.* **46**, 161–166.

Coen, C. M., and Arnold, R. W. (1972). Clay mineral genesis of some New York Spodosols. *Soil Sci. Soc. Am. Proc.* **36**, 342–350.

Damman, A. W. H. (1962). Development of hydromorphic humus Podzols and some notes on the classification of Podzols in general. *J. Soil Sci.* **13**, 92–97.

Dawson, H. J., Ugolini, F. C., Hrutfiord, B. F., and Zachara, J. (1978). Role of soluble organics in the soil processes of a Podzol, central Cascades, Washington. *Soil Sci.* **126**, 290–296.

DeConinck, F. (1980). Major mechanisms in formation of spodic horizons. *Geoderma* **24**, 101–128.

DeKimpe, C. (1970). Chemical, physical and mineralogical properties of a Podzol soil with fragipan derived from glacial till in the Province of Quebec. *Can. J. Soil Sci.* **50**, 317–330.

DeKimpe, C. R. (1974). Weathering of clay minerals in Podzols from the Appalachian Highlands. *Can J. Soil Sci.* **54**, 395–401.

DeKimpe, C. R. (1976). Influence of parent material and moisture regime on soil genesis in the Appalachian Highlands, Quebec. *Can. J. Soil Sci.* **56**, 271–283.

DeKimpe, C. R., and Martel, Y. A. (1976). Effects of vegetation on the distribution of carbon, iron, and aluminum in the B horizons of northern Appalachian Spodosols. *Soil Sci. Soc. Am. Jour.* **40**, 77–80.

DeKimpe, C. R., and McKeague, J. A. (1974). Micromorphological, physical, and chemical properties of a podzolic soil with a fragipan. *Can. J. Soil Sci.* **54**, 29–38.

DeKimpe, C. R., Baril, R. W., and Rivard, R. (1972). Characterization of a toposequence with fragipan: The Leeds-Ste. Marie-Brompton series of soils, Province of Quebec. *Can J. Soil Sci.* **52**, 135–150.

Dormaar, J. F. (1964). Humic acid associated phosphorus in some soils of Alberta. *Can. J. Soil Sci.* **43**, 235–241.

Douchafour, P., and Souchier, B. (1978). Roles of iron and clay in genesis of acid soils under a humid, temperate climate. *Geoderma* **20**, 15–26.

Evans, L. J., Rowsell, J. G., and Aspinall, J. D. (1978). Massive iron formations in some gleysolic soils of southwestern Ontario. *Can. J. Soil Sci.* **58**, 391–395.

Fedchenko, M. A. (1962). Group composition of humus in the soils of cutover areas with hairgrass in the Arkhangel'sk Oblast. *Sov. Soil Sci.,* 40-48.

Fieldes, M., and Parrott, K. W. (1966). The nature of allophane in soils: III. Rapid field and laboratory test for allophane. *New Zealand J. Sci.* **9,** 623-629.

Filimonova, L. G. (1965). Characteristics of the taiga soils of the Aldan and Tommot Rayons of the Yakut A. S. S. R. *Sov. Soil Sci.,* 219-225.

Flach, K. W., Nettleton, W. D., Gile, L. H., and Cady, J. G. (1969). Pedocementation: Induration by silica, carbonates, and sesquioxides in the Quaternary. *Soil Sci.* **107,** 442-453.

Folks, H. C., and Riecken, F. F. (1956). Physical and chemical properties of some Iowa soil profiles with clay-iron bands. *Soil Sci. Soc. Am. Proc.* **20,** 575-580.

Franzmeier, D. P., Hajek, B. F., and Simonson, C. H. (1965). Use of amorphous material to identify spodic horizons. *Soil Sci. Soc. Am. Proc.* **29,** 737-743.

Fridland, V. M. (1958). Podzolization and illimerization (clay migration). *Sov. Soil Sci.,* 24-32.

Gjems, O. (1970). Mineralogical composition and pedogenic weathering of the clay fraction in Podzol soil profiles in Zalesini, Yugoslavia. *Soil Sci.* **110,** 237-243.

Goh, K. M., and Reid, M. R. (1975). Molecular weight distribution of soil organic matter as affected by acid pre-treatment and fractionation into humic and fulvic acids. *J. Soil Sci.* **26,** 207-222.

Gorbunov, N. I. (1961). Movement of colloidal and clay particles in soils. *Sov. Soil Sci.,* 712-724.

Grossman, R. B., and Carlisle, F. J. (1969). Fragipan soils of the eastern United States. *Adv. Agron.* **21,** 237-279.

Guillet, B., Rouiller, J., and Souchier, B. (1975). Podzolization and clay migration in Spodosols of Eastern France. *Geoderma* **14,** 223-245.

Hallmark, C. T., and Smeck, N. E. (1979). The effect of extractable aluminum, iron, and silicon on strength and binding of fragipans of northeastern Ohio. *Soil Sci. Soc. Am. Jour.* **43,** 145-150.

Hansen, E. H., and Schnitzer, M. (1967). Nitric acid oxidation of Danish illuvial organic matter. *Soil Sci. Soc. Am. Proc.* **31,** 79-85.

Heilman, P. E., and Gass, C. R. (1974). Parent materials and chemical properties of mineral soils in southeast Alaska. *Soil Sci.* **117,** 21-27.

Heller, J. L. (1963). The nomenclature of soils, or what's in a name. *Soil Sci. Soc. Am Proc.* **27,** 216-220.

Hole, F. D. (1975). Some relationships between forest vegetation and Podzol B horizons in soils of Menominee tribal lands, Wisconsin, U. S. A. *Sov. Soil Sci.* **7,** 714-723.

Ignatenko, I. V. (1972). Soils of the East European forest tundra and their zonal position. *Sov. Soil Sci.* **4,** 513-526.

Karavayeva, N. A. (1973). Acid eluvial-gley soils in the middle and northern taiga of West Siberia. *Sov. Soil Sci.* **5,** 129-143.

Karavayeva, N. A., Sokolov, I. A., Sokolova, T. A., and Targul'yan, V. O. (1965). Peculiarities of soil formation in the tundra taiga frozen regions of eastern Siberia and the Far East. *Sov. Soil Sci.,* 756-766.

Karpachevskiy, L. O. (1960). Micromorphological study of leaching and podzolization of soils in a forest. *Sov. Soil Sci.,* 493-500.

Khan, S. U., and Schnitzer, M. (1971). Sephadex gel filtration of fulvic acid: The identification of major components in low-molecular weight fractions. *Soil Sci.* **112,** 231-238.

King, R. H., and Brewster, G. R. (1976). Characteristics and genesis of some subalpine Podzols (Spodosols), Banff National Park, Alberta. *Arc. Alp. Res.* **8,** 91-104.

Kodama, H., and Schnitzer, M. (1970). Kinetics and mechanism of the thermal decomposition of fulvic acid. *Soil Sci.* **109,** 265-271.

Kodama, H., and Schnitzer, M. (1971). Evidence for interlamellar adsorption of organic matter by clay in a Podzol soil. *Can. J. Soil Sci.* **51,** 509-512.

Kodama, H., and Schnitzer, M. (1973). Dissolution of chlorite minerals by fulvic acid. *Can. J. Soil Sci.* **53**, 240–243.

Kodama, H., and Schnitzer, M. (1977). Effect of fulvic acid on the crystallization of Fe(III) oxides. *Geoderma* **19**, 279–291.

Koposov, G. F. (1976). Soils of terraces in the upper course of the Taz River. *Sov. Soil Sci.* **8**, 161–172.

Kreida, N. A. (1958). Soils of the eastern European tundras. *Sov. Soil Sci.,* 51–56.

Kubota, J., and Whittig, L. D. (1960). Podzols in the vicinity of the Nelchina and Tazlina glaciers, Alaska. *Soil Sci. Soc. Am. Proc.* **24**, 133–136.

Kuz'min, V. A., and Sazonov, A. G. (1965). Podzolic soils of the Chara River basin (northern Transbaikal region). *Sov. Soil Sci.,* 1268–1276.

Laflamme, G., Baril, R., and DeKimpe, C. (1973). Caracterisation d'un Podzol Humo-Ferrique, Luvisolique, et Lithique a Esprit-Saint, Comte de Rimouski, Quebec. *Can. J. Soil Sci.* **53**, 145–154.

Lavkulich, L. M., Bhoojedhur, S., and Rowles, C. A. (1971). Soils with placic horizons on the west coast of Vancouver Island, British Columbia. *Can. J. Soil Sci.* **51**, 439–448.

Levesque, M. (1969). Characterization of model and soil organic matter metalphosphate complexes. *Can. J. Soil Sci.* **49**, 365–373.

Levesque, M., and Schnitzer, M. (1967). Organo-metallic interactions in soils: 6. Preparation and properties of fulvic acid–metal phosphates. *Soil Sci.* **103**, 183–190.

Lowe, L. E. (1975). Fractionation of acid-soluble components of soil organic matter using polyvinylpyrrolidone. *Can. J. Soil Sci.* **55**, 119–126.

Lowe, L. E. (1980). Humus fraction ratios as a means of discriminating between horizon types. *Can. J. Soil Sci.* **60**, 219–229.

Lutwick, L. E., and Dormaar, J. F. (1973). Fe status of Brunisolic and related soil profiles. *Can. J. Soil Sci.* **53**, 185–197.

MacCarthy, P., and O'Cinneide, S. (1974). Fulvic acid. II. Interaction with metal ions. *J. Soil Sci.* **25**, 429–437.

Malcolm, R. L., and McCracken, R. J. (1968). Canopy drip: A source of mobile soil organic matter for mobilization of iron and aluminum. *Soil Sci. Soc. Am. Proc.* **32**, 834–838.

McHardy, W. J., Thompson, A. P., and Goodman, B. A. (1974). Formation of iron oxides by decomposition of iron-phenolic chelates. *J. Soil Sci.* **25**, 471–482.

McKeague, J. A. (1965). Properties and genesis of three members of the Uplands catena. *Can. J. Soil Sci.* **45**, 63–77.

McKeague, J. A. (1967). An evaluation of 0.1 M pyrophosphate and pyrophosphate–dithionite in comparison with oxalate as extractants of the accumulation products in Podzols and some other soils. *Can. J. Soil Sci.* **47**, 95–99.

McKeague, J. A. (1968). Humic–fulvic acid ratio, Al, Fe, and C in pyrophosphate extracts as criteria of A and B horizons. *Can. J. Soil Sci.* **48**, 27–35.

McKeague, J. A., and Cann, D. B. (1969). Chemical and physical properties of some soils derived from reddish brown materials in the Atlantic provinces. *Can. J. Soil Sci.* **49**, 65–78.

McKeague, J. A., and Sheldrick, B. H. (1977). Sodium hydroxide-tetraborate in comparison with sodium pyrophosphate as an extractant of "complexes" characteristic of spodic horizons. *Geoderma* **19**, 97–104.

McKeague, J. A., and Sprout, P. N. (1975). Cemented subsoils (duric horizons) in some soils of British Columbia. *Can. J. Soil Sci.* **55**, 189–203.

McKeague, J. A., and St. Arnaud, R. J. (1969). Pedotranslocation: Eluviation–illuviation in soils during the Quaternary. *Soil Sci.* **107**, 428–434.

McKeague, J. A., and Wang, C. (1980). Micromorphology and energy dispersive analysis of ortstein horizons of podzolic soils from New Brunswick and Nova Scotia, Canada. *Can. J. Soil Sci.* **60**, 9–21.

McKeague, J. A., Bourbeau, G. A., and Cann, D. B. (1967). Properties and genesis of a bisequa soil from Cape Breton Island. *Can. J. Soil Sci.* **47**, 101-110. (a)

McKeague, J. A., Schnitzer, M., and Heringa, P. K. (1967). Properties of an Ironpan Humic Podzol from Newfoundland. *Can. J. Soil Sci.* **47**, 23-32. (b)

McKeague, J. A., Damman, A. W. H., and Heringa, P. K. (1968). Iron-manganese and other pans in some soils of Newfoundland. *Can. J. Soil Sci.* **48**, 243-253.

McKeague, J. A., MacDougall, J. I., Langmaid, K. K., and Bourbeau, G. A. (1969). Macro- and micromorphology of ten reddish brown soils from the Atlantic provinces. *Can. J. Soil Sci.* **49**, 53-63.

McKeague, J. A., Brydon, J. E., and Miles, N. M. (1971). Differentiation of forms of extractable iron and aluminum in soils. *Soil Sci. Soc. Am. Proc.* **35**, 33-38.

McKeague, J. A., MacDougall, J. I., and Miles, N. M. (1973). Micromorphological, physical, chemical, and mineralogical properties of a catena of soils from Prince Edward Island in relation to their classification and genesis. *Can. J. Soil Sci.* **53**, 281-295.

McKenzie, L. J., Whiteside, E. P., and Erickson, A. E. (1960). Oxidation-reduction studies on the mechanism of B horizon formation in Podzols. *Soil Sci. Soc. Am. Proc.* **24**, 300-305.

Mehuys, G. R., and DeKimpe, C. R. (1976). Saturated hydraulic conductivity in pedogenic characterization of Podzols with fragipans in Quebec. *Geoderma* **15**, 371-380.

Messenger, A. S. (1975). Climate, time, and organisms in relation to Podzol development in Michigan sands: II. Relationships between chemical element concentrations in mature tree foliage and upper humic horizons. *Soil Sci. Soc. Am. Proc.* **39**, 698-702.

Messenger, A. S., Whiteside, E. P., and Wolcott, A. R. (1972). Climate, time, and organisms in relation to Podzol development in Michigan sands: I. Site descriptions and microbiological observations. *Soil Sci. Soc. Am. Proc.* **36**, 633-638.

Mikhaylov, I. S. (1970). Main features of Chilean soils. *Sov. Soil Sci.* **2**, 1-7.

Miles, N. M., Wang, C., and McKeague, J. A. (1979). Chemical and clay mineralogical properties of ortstein soils from the Maritime provinces. *Can. J. Soil Sci.* **59**, 287-299.

Moore, T. R. (1973). The distribution of iron, manganese, and aluminum in some soils from north-east Scotland. *J. Soil Sci.* **24**, 162-171.

Moore, T. R. (1976). Sesquioxide-cemented soil horizons in northern Quebec: Their distribution, properties and genesis. *Can. J. Soil Sci.* **56**, 333-344.

Muir, A. (1961). The Podzol and podzolic soils. *Adv. Agron.* **13**, 1-56.

Muir, J. W., Morrison, R. I., Brown, C. J., and Logan, J. (1964). The mobilization of iron by aqueous extracts of plants: I. Composition of the amino-acid and organic-acid fractions of an aqueous extract of pine needles. *J. Soil Sci.* **15**, 220-225.

Naumov, Ye. M., Tsyurupa, I. C., and Andreyeva, N. A. (1964). Effect of volcanic ash on the accumulation and distribution of available compounds in the frozen-podzolic soils of Magadan Oblast. *Sov. Soil Sci.,* 1299-1309.

Nicholson, H. M., and Moore, T. R. (1977). Pedogenesis in a subarctic iron-rich environment, Schefferville, Quebec. *Can. J. Soil Sci.* **57**, 35-45.

Nikitin, Ye. D., and Fedorov, K. N. (1977). Composition and genesis of the ortsands of soils in the taiga zone of West Siberia. *Sov. Soil Sci.* **9**, 383-391.

Nikonov, V. V. (1979). Characteristics of soil formation in the northern taiga spruce biogeocenoses of the Kola Peninsula. *Sov. Soil Sci.* **11**, 501-511.

Nornberg, P. (1980). Mineralogy of a Podzol formed in sandy materials in northern Denmark. *Geoderma* **24**, 25-43.

Parfenova, Ye. I., and Yarilova, Ye. A. (1960). The problem of lessivage and podzolization. *Sov. Soil Sci.,* 913-925.

Pawluk, S. (1960). Some Podzol soils of Alberta. *Can. J. Soil Sci.* **40**, 1-14.

Pawluk, S. (1972). Measurement of crystalline and amorphous iron removal in soils. *Can. J. Soil Sci.* **52**, 119–123.

Pedro, G., Jamagne, M., and Begon, J. C. (1978). Two routes in genesis of strongly differentiated acid soils under humid, cool-temperate conditions. *Geoderma* **20**, 173–189.

Pollack, S. S., Lentz, H., and Ziechmann, W. (1971). X-ray diffraction study of humic acids. *Soil Sci.* **112**, 318–324.

Proudfoot, V. B. (1958). Problems of soil history. Podzol development at Goodland and Torr Townlands, Co. Antrim, Northern Ireland. *J. Soil Sci.* **9**, 186–198.

Pshenichnikov, B. F. (1974). Podzolic Al–Fe–humus soils of southern Sakhalin *Sov. Soil Sci.* **6**, 263–268.

Pshenichnikov, B. F. (1976). Podzolic illuvial-humic soils of Primorskiy Kray. *Sov. Soil Sci.* **8**, 632–640.

Ragg, J. M., and Ball, D. F. (1964). Soils of the ultra-basic rocks of the Island of Rhum. *J. Soil Sci.* **15**, 124–133.

Ragg, J. M., Bracewell, J. M., Logan, J., and Robertson, L. (1978). Some characteristics of the Brown Forest soils of Scotland. *J. Soil Sci.* **29**, 228–242.

Raman, K. V., and Mortland, M. M. (1969). Amorphous materials in a Spodosol: Some mineralogical and chemical properties. *Geoderma* **3**, 37–43.

Rieger, S. (1974). Humods in relation to volcanic ash in southern Alaska. *Soil Sci. Soc. Am. Proc.* **38**, 347–351.

Rieger, S., and DeMent, J. A. (1965). Cryorthods of the Cook Inlet–Susitna lowland, Alaska. *Soil Sci. Soc. Am. Proc.* **29**, 448–453.

Rieger, S., Schoephorster, D. B., and Furbush, C. E. (1979). "Exploratory Soil Survey of Alaska." U. S. Dept. Agr., Soil Cons. Serv. Washington, D. C.

Rode, A. A. (1937). "Podzol-forming Process." U. S. Dept Comm., Trans. TT 69–55088, Springfield, Virginia.

Romans, J. C. C. (1962). The origin of the indurated B3 horizon of podzolic soils in north-east Scotland. *J. Soil Sci.* **13**, 141–147.

Romans, J. C. C., Stevens, J. H., and Robertson, L. (1966). Alpine soils of north-east Scotland. *J. Soil Sci.* **17**, 184–199.

Rozhnova, I. A., and Schastnaia, L. S. (1959). Study of the interrelationships of vegetation and soil in the Karelian Isthmus. *Sov. Soil Sci.,* 15–24.

Rozov, N. N., and Ivanova, Ye. N. (1967). Classification of the soils of the U. S. S. R. (Principles and a systematic list of soil groups). *Sov. Soil Sci.,* 147–155.

Rubilin, Ye. V., and Dzhumagulov, M. (1977). Humus and its composition in some subalpine and alpine soils of Kirghizia. *Sov. Soil Sci.* **9**, 264–272.

Rubtsov, D. M. (1964). Gley weakly-podzolic soils. *Sov. Soil Sci.,* 677–681.

Rusanova, G. V. (1979). Soil genesis on binary parent materials. *Sov. Soil Sci.* **11**, 274–284.

Saini, G. R., MacLean, A. A., and Doyle, J. J. (1966). The influence of some physical and chemical properties on soil aggregation and response to VAMA. *Can. J. Soil Sci.* **46**, 155–160.

Schnitzer, M. (1969). Reactions between fulvic acid, a soil humic compound and inorganic soil constituents. *Soil Sci. Soc. Am. Proc.* **33**, 75–81.

Schnitzer, M. (1970). Characteristics of organic matter extracted from Podzol B horizons. *Can. J. Soil Sci.* **50**, 199–204.

Schnitzer, M., and DeLong, W. A. (1955). Investigations on the mobilization and transport of iron in forested soils. II. The nature of the reaction of leaf extracts and leachates with iron. *Soil Sci. Soc Am Proc.* **19**, 363–368.

Schnitzer, M., and Desjardins, J. G. (1962). Molecular and equivalent weights of the organic matter of a Podzol. *Soil Sci. Soc. Am. Proc.* **26**, 362–365.

Schnitzer, M., and Desjardins, J. G. (1969). Chemical characteristics of a natural soil leachate from a humic Podzol. *Can. J. Soil Sci.* **49**, 151-158.

Schnitzer, M., and Hansen, E. H. (1970). Organo-metallic interactions in soils. 8. An evaluation of methods for the determination of stability constants of metal–fulvic acid complexes. *Soil Sci.* **109**, 333-340.

Schnitzer, M., and Kodama, H. (1967). Reactions between a Podzol fulvic acid and Na-montmorillonite. *Soil Sci. Soc. Am. Proc.* **31**, 632-636.

Schnitzer, M., and Kodama, H. (1976). The dissolution of micas by fulvic acid. *Geoderma* **15**, 381-391.

Schnitzer, M., and Skinner, S. I. M. (1963). Organo-metallic interactions in soils. 2. Reactions between different forms of iron and aluminum and the organic matter of a Podzol Bh horizon. *Soil Sci.* **96**, 181-186.

Schnitzer, M., and Skinner, S. I. M. (1964). Organo-metallic interactions in soils. 3. Properties of iron- and aluminum-organic matter complexes, prepared in the laboratory and extracted from a soil. *Soil Sci.* **98**, 197-203.

Schnitzer, M., and Skinner, S. I. M. (1965). Organo-metallic interactions in soils. 4. Carboxyl and hydroxyl groups in organic matter and metal retention. *Soil Sci.* **99**, 278-284.

Schnitzer, M., and Skinner, S. I. M. (1968). Alkali versus acid extraction of soil organic matter. *Soil Sci.* **105**, 392-396.

Schnitzer, M., and Vendette, E. (1975). Chemistry of humic substances extracted from an arctic soil. *Can. J. Soil Sci.* **55**, 93-103.

Singer, M., Ugolini, F. C., and Zachara, J. (1978). In situ study of podzolization on tephra and bedrock. *Soil Sci. Soc. Am. Jour.* **42**, 105-111.

Sneddon, J. I., Lavkulich, L. M., and Farstad, L. (1972). The morphology and genesis of some alpine soils in British Columbia, Canada: I. Morphology, classification, and genesis. *Soil Sci. Soc. Am Proc.* **36**, 100-104. (a)

Sneddon, J. I., Lavkulich, L. M., and Farstad, L. (1972). The morphology and genesis of some alpine soils in British Columbia, Canada: II. Physical, chemical, and mineralogical determinations and genesis. *Soil Sci. Soc. Am. Proc.* **36**, 104-110. (b)

Sokolov, I. A., Gradusov, B. P., Tursina, T. V., Tsyurupa, I. G., and Tyapkina, N. A. (1974). Description of soil formation on unconsolidated silicate rocks in the permafrost-taiga region. *Sov. Soil Sci.* **6**, 269-282.

Sokolova, T. A. (1964). Effect of rocks on Podzol formation. *Sov. Soil Sci.*, 233-240.

Sokolova, T. A., Targul'yan, V. O., and Smirnova, G. Ya. (1971). Clay minerals in Fe-Al-humus podzolic soils and their role in the formation of the soil profile. *Sov. Soil Sci.* **3**, 331-341.

Stephens, F. R. (1969). A forest ecosystem on a glacier in Alaska. *Arctic* **22**, 441-444.

Stevens, J. H., and Wilson, M. J. (1970). Alpine Podzol soil on the Ben Lawers massif, Perthshire. *J. Soil Sci.* **21**, 85-95.

Stobbe, P. C., and Wright, J. R. (1959). Modern concepts of the genesis of Podzols. *Soil Sci. Soc. Am. Proc.* **23**, 161-164.

Targul'yan, V. O. (1964). Movement of suspensions in mountain-tundra and mountain-taiga soils on massive-crystalline rocks. *Sov. Soil Sci.*, 800-810.

Tonkonogov, V. D. (1970). Genesis of manganese–iron new formations in sandy Podzols. *Sov. Soil Sci.* **2**, 159-167.

Tonkonogov, V. D. (1971). Experimental statistical analysis of geographic soil formation patterns, using the sandy Podzols of the northern part of the Russian plain as an example. *Sov. Soil Sci.* **3**, 29-39.

Ufimtseva, K. A. (1963). Mountain taiga soils of the Transbaikal region. *Sov. Soil Sci.*, 241-248.

Ugolini, F. C. (1966). Soils of the Mesters Vig District, northeast Greenland: I. The Arctic Brown and related soils. *Medd. om Gron.*, Bd. **176**, No. 1, 22 pp.

Ugolini, F. C., Minden, R., Dawson, J., and Zachara, J. (1977). An example of soil processes in the *Abies amabilis* zone of central Cascades, Washington. *Soil Sci.* **124**, 291–302.

Valentine, K. W. G. (1969). A Placic Humic Podzol on Vancouver Island, British Columbia. *Can. J. Soil Sci.* **49**, 411–413.

Wada, K., and Harward, M. E. (1974). Amorphous clay constituents of soils. *Adv. Agron.* **26**, 211–260.

Wang, C., and Rees, H. W. (1980). Characteristics and classification of non-cemented sandy soils in New Brunswick. *Can. J. Soil Sci.* **60**, 71–81.

Wang, C., Nowland, J. L., and Kodama, H. (1974). Properties of two fragipan soils in Nova Scotia including scanning electron micrographs. *Can. J. Soil Sci.* **54**, 159–170.

Wicklund, R. F., and Whiteside, E. P. (1959). Morphology and genesis of the soils of the Caribou catena in New Brunswick, Canada. *Can. J. Soil Sci.* **39**, 222–234.

Wright, J. R., and Schnitzer, M. (1963). Metallic–organic interactions associated with podzolization. *Soil Sci. Soc. Am. Proc.* **27**, 171–176.

Wurman, E., Whiteside, E. P., and Mortland, M. M. (1959). Properties and genesis of finer textured subsoil bands in some sandy Michigan soils. *Soil Sci. Soc. Am. Proc.* **23**, 135–143.

Yassoglou, N. J., and Whiteside, E. P. (1960). Morphology and genesis of some soils containing fragipans in northern Michigan. *Soil Sci. Soc. Am. Proc.* **24**, 396–407.

Zaboyeva, I. V. (1965). Gley-podzolic soils of northeast European U. S. S. R. *Sov. Soil Sci.*, 745–755.

Chapter 6

ALFISOLS

Alfisols are soils in which the dominant genetic process has been the translocation in aqueous suspension of whole fine clay from the upper mineral horizon of the soil to a lower horizon with little or no concurrent translocation of organic matter. The typical profile developed by this process in cold regions consists of a fairly thin layer of partially decomposed organic litter (an 0 horizon), a thin, dark gray A horizon, a gray or grayish brown horizon commonly with platy structure (an albic horizon), a brown or yellowish brown horizon of clay accumulation with blocky or prismatic structure (an argillic horizon), and a C horizon enriched in calcium and other carbonates. The A horizon is absent in some Alfisols. Tonguing or interfingering of the albic horizon in the argillic horizon is common. Alfisols do not have a mollic epipedon or a spodic horizon.

Most cold Alfisols are in the warmer parts of the boreal belt under a mixed forest that tends to produce a ground litter that is high in bases, but they also occur under coniferous forest especially in mountainous areas, under steppe vegetation, and, in a few places, under tundra. Some are underlain by permafrost at depths great enough that perched water and cryoturbation do not interfere with the soil-forming process. They exist under a wide range of rainfall regimes, from subhumid continental to humid (but not perhumid) maritime. Most have loamy or clayey textures, but some are sandy. In every case they occur in landscapes that have been stable for a long period of time.

I. Genesis of Alfisols

A. CONDITIONS OF FORMATION

Conditions that favor the development of Alfisols are *(a)* loamy or clayey parent materials; *(b)* high base saturation and weakly acidic to neutral reaction; *(c)* short, warm summers that stimulate intense microbial activity in the 0 horizon; *(d)* alternating wet and dry conditions in the soil; *(e)* slow percolation rates and limited depth of penetration of rainwater but enough precipitation to saturate and leach the upper horizons of the soil periodically (McKeague and St. Arnaud, 1969; Sokolov *et al.,* 1974; Pedro *et al.,* 1978). The existence of any one of these conditions is generally sufficient to inhibit the operation of the spodic process, but in very humid areas and in coarse materials, a strongly leaching regimen can be established despite otherwise adverse conditions, and Spodosols rather than Alfisols will form. In many humid and subhumid areas, Spodosols form after development of an Alfisol has modified the parent material to the extent that the spodic process can proceed.

B. RELATIONSHIP TO "PODZOLS"

Most Alfisols were described as podzolic soils in early soil taxonomies because of the presence of an albic, or podzolic, horizon in the soil profile. They are still considered to be podzolic soils in the official U.S.S.R. soil classification system (Rozov and Ivanova, 1967). In recent years, however, illuviation of whole silicate clays has increasingly been recognized as a separate and distinct process from that involving illuviation of organometallic complexes, or "podzolization" (Pedro *et al.,* 1978). Several names have been proposed for this process, including *silication* (Bray, 1934), *lessivage* (Duchafour, 1951), *illimerization* (Fridland, 1958), and *argilluviation* (Dudal, 1968). The latter name has been adopted here.

C. ARGILLUVIATION

Particles involved in argilluviation are mainly finer than .2 microns $(.2\mu)$—that is, in the fine clay-size fraction (Pawluk, 1961; Floate, 1966; Stonehouse and St. Arnaud, 1971). Because montmorillonite and other smectites are dominant in this fraction in cold soils, they are the principal migrating silicate clays (Gradusov and Dyazdevich, 1961; Floate, 1966; Sokolov *et al.,* 1974). In strong contrast with the Spodosols, in which smectites occur almost universally in the albic horizon but are seldom observed in the spodic horizon, the proportion of smectites is greater in the argillic horizon of Alfisols than in the overlying albic horizon. In older Alfisols smectites may be virtually absent in the albic horizon but dominant in the fine clay fraction of the argillic and lower horizons (Pawluk, 1961).

Clay translocation generally takes place in soils with a weakly acidic to neutral reaction (Frei, 1967; Pedro *et al.,* 1978). At higher pH values salts, including free calcium carbonate, tend to flocculate clays and prevent them from entering into suspension. At pH values below 5, clays flocculate because of the high hydrogen ion concentration in the soil solution (Sokolov *et al.,* 1974). Also, under very strongly acidic conditions, organic acids adsorbed on clay faces or in interlamellar spaces tend to inhibit movement and to attack and break down the lattice structure. Nevertheless, where the clay content of the parent material is high, clay migration will occur even under strongly acidic conditions.

In calcareous parent materials the first stage of soil development normally is the accumulation of organic matter in the upper part of the mineral soil to form either a mollic epipedon or a horizon that satisfies the requirements for a mollic epipedon except for thickness. Because of the large supply of bases, microbial activity rapidly breaks down organic matter in the 0 horizon into water and carbon dioxide (McKeague and St. Arnaud, 1969), and there is only minimal production of strongly acidic organic compounds. The ratio of fulvic acid to humic acid is much lower than in acidic soils and, as a consequence, the rate of weathering and decomposition of minerals is lower (Zonn, 1973). In time, if precipitation is high enough that leaching of bases exceeds the rate of replenishment by biological recycling and weathering of primary minerals, excess bases are removed from the upper soil horizon although base saturation of the exchange complex remains high. With the removal of unadsorbed bases from the soil solution, the rate of organic matter accumulation in the A horizon decreases, and eventually the organic matter content of the horizon is reduced. When the pH drops to or below neutrality and the flocculating effect of salts and organic matter is diminished, dispersion of clays occurs after rains, and illuviation can begin (Wright *et al.,* 1959; Pettapiece and Zwarich, 1970; McKeague *et al.,* 1973). In parent materials that are not calcareous, this stage in the process is attained more quickly and may be the beginning stage.

No chemical or physical changes occur in the fine clays as they are carried downward by percolating water. They are removed from suspension and deposited either where percolation slows or stops (Buol and Hole, 1961; McKeague and St. Arnaud, 1969; Sokolov *et al.,* 1974) or, in calcareous or other basic materials, when they reach a layer with a high concentration of calcium or other cations that causes them to coagulate (Frei, 1967). In deep loessial soils of northwestern U.S.S.R., for example, the lower boundary of deposition may be 150–200 cm in noncalcareous materials but only 50–75 cm in calcareous materials (Zonn, 1973). In soils with a texturally contrasting substratum such as a clayey moraine deposit under a mantle of loess—a common occurrence in glaciated areas—percolation is arrested at the interface between the two parent materials, and the suspended clay is deposited immediately below it (McKeague and Cann, 1969; Zonn, 1973).

The initial deposit of suspended fine clay may be as a lining on the walls of larger pores, as a series of thin, roughly horizontal lamellae or, in the coarsest soils, as a coating on sand grains or coarse fragments (Karpachevskiy, 1960). In loamy and clayey soils, the clay may be dispersed from the initial deposits by transport in capillaries (Soil Survey Staff, 1975), by internal soil movements generated by freezing and thawing and alternate wetting and drying, by slow mass movement, or by the activities of organisms (Buol and Hole, 1961). The increased content of swelling clays in the incipient argillic horizon results in greater volume changes on wetting and drying, and this in turn eventually leads to the formation of the blocky structure that is characteristic of argillic horizons in loamy soils (Kremer, 1969; McKeague et al., 1972). Clay that migrates to the argillic horizon after development of the structure is deposited in very fine increments on the surfaces of the peds. This produces a laminated coating of fine clays in which the platelike particles are oriented parallel to the ped surface. These clay skins (argillans, clay cutans) can usually be identified by their birefringence and uniform or wavy extinction when examined under a polarizing microscope. It is possible to confuse clay skins formed by illuviation with stress cutans that result from shrinking and swelling or from freezing pressures (McKeague and St. Arnaud, 1969), but illuvial clay skins commonly exhibit flow lobes whereas stress cutans do not. The flow pattern, in many cases, can be observed with the aid of a hand lens in the field. In coarse soils the illuvial clay tends to form bridges between coated sand grains that also can be seen under a lens or microscope.

As leaching continues and the upper horizons lose bases and become more acidic, increasing amounts of iron and aluminum either go into solution or migrate as crystalline hydrous oxides of fine clay size in suspension (Gradusov and Dyazdevich, 1961; Pawluk, 1961). Some oxides are adsorbed on the clays and migrate with them (Harlan et al., 1977), but in general the silicate clays and the oxides move independently (Zonn, 1973). In older soils small quantities of iron and aluminum may be translocated in the form of organometallic complexes (McKeague et al., 1973; Sokolov et al., 1974; Pawluk and Dudas; 1978). In some soils of humid areas, organometallic complexes are deposited in the upper part of the argillic horizon in sufficient quantity to approach or exceed the requirements for a spodic horizon (McKeague et al., 1971). Other elements, including manganese and phosphorus, also move downward with the percolating water (Buol and Hole, 1961). All are deposited, along with the fine clay, in the argillic horizon.

The result of these migrations is the development of an albic horizon that is depleted of free iron and aluminum and is strongly acid, and an argillic horizon that is higher in iron and aluminum than the parent material (Sokolov et al., 1976; Pedro et al., 1978) and is less acidic than the albic horizon (Fig. 21). In most cold Alfisols the fulvic acid fraction in the argillic horizon is larger than the humic acid fraction, but the phenolic acid component of the fulvic acid is small,

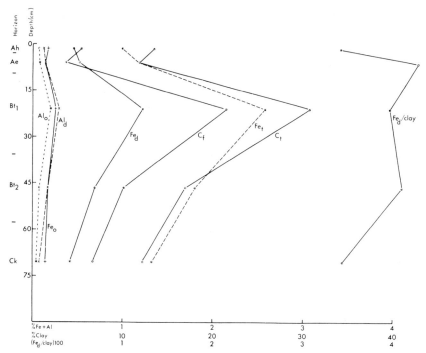

FIGURE 21. *Distribution of total iron (Fe_t), oxalate-extractable iron (Fe_o) and aluminum (Al_o), dithionite-extractable iron (Fe_d) and aluminum (Al_d), total ($< .002$ mm) clay (C_t), and fine ($< .0002$ mm) clay (C_f) in an Alfisol from Saskatchewan. (From Stonehouse and St. Arnaud [1971] by permission of the Agricultural Institute of Canada.)*

and the total amount of organic matter that accumulates in the illuvial horizon is very low (Lowe, 1980). In these respects, as well as in the accumulation of unaltered fine clay, the argillic horizon differs significantly from the spodic horizon.

When development has proceeded to the point that there is no longer a substantial source of clay in the albic horizon, percolating waters take up the clay in the upper part of the argillic horizon and redeposit it deeper in the profile. A residue of light gray, sandy and silty material identical with that in the albic horizon is left behind and forms a second layer above the remaining clay skins on peds, the walls of pores, and the borders of desiccation cracks. Eventually the clay skins are replaced completely. In well-drained loamy soils, the effects of this process are a continually thickening albic horizon, an intermediate horizon in which clay skins are partially removed and fingers or tongues of albic material alternate with remnants of the original argillic horizon, and a deeper horizon in which newly formed clay skins are thick and continuous over structural units (Zonn, 1966; Kremer, 1969; Ranney and

Beatty, 1969; Pettapiece and Zwarich, 1970; Rusanova, 1976) (Fig. 22). Deep tongues of albic material carried in by percolating water may extend in desiccation cracks to depths as great as 1.5 m (Fig. 23).

D. ONSET OF SPODIC PROCESS

A final stage in the development of well-drained soils with argillic horizons in cold, humid or subhumid areas occurs after the albic horizon is depleted of fine clay and bases and is strongly acid. Conditions are then suitable for operation of the spodic process, and a spodic horizon may form within the albic horizon of the original Alfisol. Soils in which this has taken place are classified as Boralfic Cryorthods (Chapter 5). In some cases the superimposed spodic horizon may form below the albic horizon and replace, at least in part, the earlier argillic horizon (Pedro *et al.*, 1978).

E. REMNANT MOLLIC EPIPEDONS

In the southern part of the boreal belt, at the boundary between the boreal forest and the grasslands of semiarid areas, a gradual encroachment of forests into former grasslands has taken place as a result of climatic change since the warm period (the hypsithermal or climatic optimum) that existed 4000 to 7000 years ago. Coincident with this climatic change and the greater leaching efficiency that accompanied it, the mollic epipedon that had developed under steppe or mixed grass and forest vegetation was subjected to acidification and degradation (Sawyer and Pawluk, 1963; Kuz'min, 1969; Pettapiece and Zwarich, 1970). Increasing dominance of the argilluviation process resulted in the formation of an argillic horizon (or the enhancement of a previously existing argillic horizon) and in the development of an albic horizon in place of the original mollic epipedon. Remnants of the original epipedon remain in many places, however, commonly in the form of dark bands or spots in the lower part of the albic and the upper part of the argillic horizons. They are best preserved in calcareous, fine-textured soils with slow internal drainage. Carbon-14 dating of organic matter in these remnant horizons demonstrates that they are 6000 to 7000 years old, whereas organic matter in the present surface horizons is less than 1300 years old. Also, the remnant organic matter is very high in humic acid, which is characteristic of comparatively warm soils of grasslands, and the surface horizons have a high proportion of fulvic acid, a

---

FIGURE 22. *Distribution of oriented clay skins in segments of horizons in an Alfisol from Wisconsin. Actual colors of clay skins are yellowish and reddish brown. Clay skins in deeper horizons occur principally on ped surfaces; those in the B horizons tend to be in the interior of peds (Buol and Hole, 1961). (Reprinted from* Soil Science Society of America Proceedings, *Volume 25, page 378, 1961. By permission of the Soil Science Society of America.)*

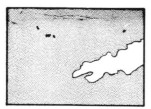 View from A_2
horizon(5.5"–10")
(av. 0.03% clay-
skin, by vol.)

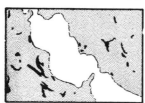 View from B_1
horizon(10"–19")
(av. 0.69% clay-
skin, by vol.)

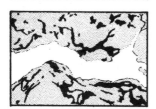 View from B_2
horizon (19"–31")
(av. 2.69% clay-
skin, by vol.)

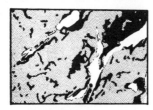 View from B_3
horizon(31"–45")
(av. 3.15% clay-
skin, by vol.)

 View from C_1
horizon(45"–59")
(av. 5.27% clay-
skin, by vol.)

 View from C_2
horizon(59"–69")
(av. 2.05% clay-
skin, by vol.)

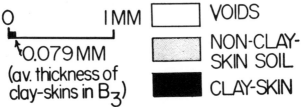

0   1 MM   ☐ VOIDS

↑0.079 MM   ▨ NON-CLAY-
(av. thickness of   SKIN SOIL
clay-skins in B_3)   ■ CLAY-SKIN

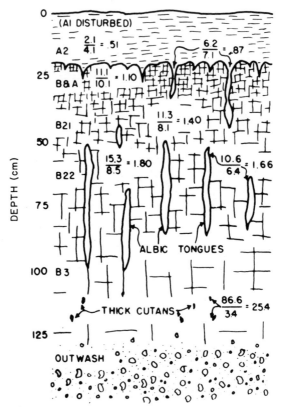

FIGURE 23. *Deep tongues of albic material and accumulation of fine clay in cutans (clay skins) in deeper horizons in a northern Wisconsin Alfisol. The numerator in the fractions is the percentage of fine clay (< .0002 mm), and the denominator is the percentage of coarse clay (.002–.0002 mm). The ratio between the two is shown to the right (Ranney and Beatty, 1969).* (Reprinted from Soil Science Society of America Proceedings, Volume 33, page 771, 1969. By permission of the Soil Science Society of America.)

condition that is normally associated with cold forest soils (Dobrovol'skiy *et al.*, 1970; Tolchel'nikov, 1970; Ufimtseva, 1970).[*]

F. OVERSATURATION

The argillic horizon of older soils, especially those forming in clayey materials, may become very slowly permeable. After heavy rains moisture may perch or move laterally in the albic horizon above it. The albic horizon and, in some cases, the upper part of the argillic horizon in these soils experience

[*] Soils of this kind are transitional between Mollisols and Alfisols. If the combined present surface mineral horizon and the remnant of the earlier epipedon, although separated by an albic horizon, still meet the requirements for a mollic epipedon, the soil is classified as a Mollisol.

periodic reducing conditions when the soil is saturated and oxidizing conditions in dry periods (Fig. 24). During periods of saturation iron is converted to the ferrous state and some of it removed in solution by water moving laterally above the argillic horizon (Pedro *et al.,* 1978), but much of it is deposited as streaks and mottles or in iron–manganese concretions on subsequent drying (Zonn, 1966; Zonn, 1973). Albic horizons that are subject to periodic saturation are commonly strongly acid (Zaboyeva, 1958; Rybtsov, 1960). In areas with very heavy precipitation concentrated in a short wet season, as in extreme southeastern Siberia and eastern Manchuria, they may become very thick (Liverovskiy and Roslikova, 1962). The transition between the albic and argillic horizons in soils with slowly permeable argillic horizons is commonly abrupt and smooth (Zaboyeva, 1965), but in places tongues of the albic horizon may extend into the argillic horizon (Ufimtseva, 1970; Rusanova, 1976).

Concretions that form in the lower part of the albic horizon and the upper part of the argillic horizon consist principally of iron and manganese but also contain magnesium, aluminum, phosphorus, and organic matter (Rusanova *et al.,* 1975). Concretion formation is probably more a biological than a strictly chemical process (Polteva and Sokolova, 1967). It is likely that most of them form in the late summer when the soil is drying and atmospheric air in pores is in contact with the still-saturated soil matrix, a situation that is most favorable for the growth of iron- and manganese-fixing bacteria (Aristovskaya, 1963).

In areas where freezing is deep and the soils are not completely thawed until late summer, water may perch above the frozen zone even in soils that do not have a slowly permeable illuvial horizon. This leads to reducing conditions and

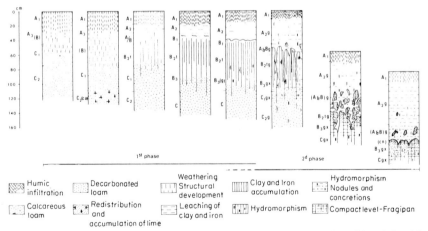

FIGURE 24. *Stages in the development of an argillic horizon, subsequent oversaturation and degradation of the horizon, and formation of a compact deep horizon with characteristics of a fragipan. Few, if any, Alfisols of cold regions reach this final stage. (From G. Pedro, M. Jamagne, and J. C. Begon [1978]. Two routes in genesis of strongly differentiated acid soils under humid, cool-temperate conditions.* Geoderma 20, *173-189.)*

eventual mottling in the albic horizon. At the same time moisture accumulates below the frozen zone and gleying takes place there as well. Most of the argillic horizon, which remains frozen longest, shows no sign of excessive wetness (Ovchinnikov et al., 1973).

II. Classification of Cold Alfisols

Most well- and moderately well-drained Alfisols of cold regions are in the great group of Cryoboralfs. Those that have developed in saline materials and have a natric horizon in place of the argillic horizon are Natriboralfs. Poorly drained Alfisols are Aqualfs.

A. BORALFS

Typic Cryoboralfs are those in which an argillic horizon of loamy or clayey texture exists but in which development has not progressed to the point that either deep tongues of albic material project into the argillic horizon or the argillic horizon has become so slowly permeable that the albic horizon above it is saturated for lengthy periods after rains. Albic coatings on peds (interfingering) are common, however. The soils either have no A horizon below a thin 0 horizon or have a very thin A horizon. The albic horizon generally is moderately thick, has platy structure, and is unmottled. The argillic horizon is dark brown or dark yellowish brown and has blocky structure. In humid areas and in areas where the soils are formed in clayey acidic materials, the albic horizon may be strongly acid (Pawluk, 1961; Polteva and Sokolova, 1967; McKeague et al., 1972; Pawluk and Dudas, 1978). In general, however, the albic horizon is moderately to slightly acid and the argillic horizon is slightly acid to neutral. The underlying material may or may not be calcareous.

Typic Cryoboralfs occur principally near the warm limit of the boreal forest close to its border with the grasslands of less humid areas (Williams and Bowser, 1952; Kuz'min, 1973), but they also form in fine-grained materials of colder areas (Ufimtseva, 1968) and in neutral to alkaline materials of humid maritime areas (McKeague et al., 1972). The vegetation is most commonly a mixed forest but may be a larch forest (Sokolov et al., 1974), fir–spruce forest (Chernov, 1965), or mixed pine and steppe (Makeyev and Nogina, 1958). In the humid areas they are associated with Spodosols in coarser or more acidic materials (Stobbe, 1952). Many have characteristics approaching those of Spodosols, such as reddish brown colors and some accumulation of organic matter in the argillic horizon (McKeague et al., 1971, 1973). In drier continental areas they grade to Mollisols in grasslands or in younger or more calcareous materials. Where they occur well into the boreal forest belt, they are associated with Cryochrepts in younger or calcareous loamy materials, with Cryorthods in sands, and with Histosols or Cryaquepts over permafrost in depressions (Leahey, 1947; Sokolov *et al.,* 1974). Where parent materials are high in sodium and other soluble salts, the Cryoboralfs are in higher areas, and Natriboralfs and other saline soils occur in depressions (Yelovskaya, 1965; Floate, 1966).

Glossic Cryoboralfs have tongues of albic material extending well into the argillic horizon and exhibit considerable degradation of clay skins in the upper part of the horizon. In the lower part of the argillic horizon, clay skins are thick on ped surfaces, and the albic material is confined to large vertical fissures (Rusanova, 1976). Much of that material is silt and sand that has sifted into the fissures during periods of dryness rather than a residue after clay removal (Ranney and Beatty, 1969). The albic horizon in these older soils may be extremely acid, but the argillic horizon is generally only moderately acidic (Polteva and Sokolova, 1967).

Glossic Cryoboralfs occur in areas where the annual precipitation exceeds 400 mm, but in which there are prolonged relatively dry periods in summer. They are mostly on mountain slopes and on moraine plains. On the plains, as in the northwestern U.S.S.R., they are associated with Cryaqualfs and Histosols in depressions (Rusanova *et al.,* 1975) and grade to Cryaqualfs in upland positions to the north. Soils in sandy material under the same climatic conditions are Spodosols.

Mollic Cryoboralfs are soils in which a mollic epipedon has degraded as a result of climatic change and invasion of grassland by forest but which still retain some of the organic matter and dark colors of the former epipedon, either as a remnant at the bottom of the albic horizon or interspersed with bleached sand grains in the upper part of the soil. Tongues of albic material that extend into the argillic horizon in some of these soils probably are relict features from a still earlier climatic episode. Soils of this kind occur in continental areas at the border between the boreal forest and grasslands, where there is a history of alternate advance and retreat of the forest (Jungerius, 1969).

Psammentic Cryoboralfs develop in sandy materials. The argillic horizon in these soils may consist of clay bridges between coated sand grains or may be a series of roughly horizontal bands, or lamellae, whose combined thickness is great enough to satisfy the minimum requirement for the argillic horizon. The albic horizon may be thicker than 60 cm (Soil Survey Staff, 1975).

Andeptic Cryoboralfs have a surface mantle of volcanic ash or are formed entirely in ash. They are known to occur in the northern Rocky Mountains in areas with moderate precipitation. The ash weathers readily to amorphous aluminosilicates that, under alternately wet and dry conditions, combine with magnesium and iron to form smectites (Chichester *et al.,* 1969). These, in turn, are subject to intermittent leaching and accumulation in an argillic horizon.

Most of the clay in the argillic horizons of such soils, however, is formed in place and is simply augmented by illuvial clay (Pettapiece and Pawluk, 1972). The presence of an albic horizon that may have tongues extending into the argillic horizon distinguishes these soils from the Andepts, in which no clay translocation has taken place. Under higher and more uniform precipitation, a spodic rather than an argillic illuvial horizon is more likely to develop in ashy soils.

Pergelic Cryoboralfs are well-drained soils underlain by permafrost at a depth great enough that it does not interfere with the argilluviation process. The soils occur under a continental climate with low snowfall and heavy summer showers. Under these conditions most of the snow disappears before the soil begins to thaw in the spring, and moisture from summer rains does not penetrate as far as the permafrost table. As a result there is little or no cryoturbation and no physical disruption of the argillic horizon. The soils are associated with Pergelic Cryochrepts on younger or more highly calcareous material and with Cryaquepts and Histosols with shallow permafrost tables in depressions (Sokolov *et al.*, 1974). In very dry areas soils in the depressions may be saline (Yelovskaya, 1965). No subgroup of Pergelic Cryoboralfs is currently listed in the United States Taxonomy, but soils of this kind have been observed in eastern Siberia (Filimonova, 1965; Sokolov *et al.*, 1969; Sokolov *et al.*, 1976) and northwestern Canada (Pettapiece, 1975).

Aquic Cryoboralfs are soils in which moisture from precipitation perches above the argillic horizon for periods long enough to produce periodic reducing conditions in the albic horizon and, in some cases, in the upper part of the argillic horizon. The soils occupy upland segments of the landscape and have deep groundwater tables. Perching takes place either because the argillic horizon is very high in clay and is slowly permeable or because it remains frozen well into the summer. Most of these soils support coniferous forests. Mottling, concretions, and other characteristics of wetness are typical of the upper part of the soil, but most of the argillic horizon has brown colors with no mottling. Because of the frequent alternation between oxidizing and reducing conditions in the albic horizon there is even less translocation of iron and organometallic complexes than in other Cryoboralfs (McKeague *et al.*, 1973).

Aquic Cryoboralfs occur in places in the warmer parts of the boreal belt (Stonehouse and St. Arnaud, 1971), but they are dominant only in fine-grained parent materials in humid areas adjacent to the tundra (Rybtsov, 1960; Zaboyeva, 1965). In those areas the thickness of the albic horizon decreases, and argilluviation becomes less pronounced as the boundary with the polar belt is approached and the surface vegetation becomes increasingly mossy (Zaboyeva, 1958). They grade to poorly drained Cryaquepts with shallow permafrost in tundra areas and to Typic and Glossic Cryoboralfs in the direction of a warmer climate. In associated sandy materials of uplands, most soils are Spodosols.

Natriboralfs are moderately well-drained Alfisols with a natric rather than an argillic horizon. They form in fine-grained parent materials that are high in exchangeable sodium and magnesium and relatively low in exchangeable calcium (Clark and Green, 1964; Reeder *et al.*, 1967). They normally occur in shallow depressions or at the edges of deeper depressions and valleys in areas with subhumid or semiarid continental climates (Floate, 1966). Most of them support only grass or shrub vegetation, although associated nonsaline Cryoboralfs, Cryoborolls, or Cryochrepts are forested. Saline or alkali soils with salt crusts on the surface may occupy adjacent poorly drained depressions (Yelovskaya, 1965; Pringle *et al.*, 1975).

Salinity in these soils is generally caused by groundwater high in sodium or magnesium carbonates. The salts move toward the surface during droughty periods in midsummer or during freezing in autumn and are flushed out of the solum after a rain or flood (Yelovskaya, 1965; Ufimtseva, 1968). Under these conditions large amounts of fine clay, dominantly montmorillonite, are translocated into the illuvial horizon even under relatively low atmospheric precipitation rates (Floate, 1966). This results in a greater degree of swelling and shrinking on wetting and drying, and in the development of prismatic or columnar structure in the natric horizon. A blocky secondary structure may form. In soils that are high in sodium, dispersed humus lines the sides of the prisms. Organic matter in both the A horizons and the prism linings in these soils consists principally of humic acid (Anderson *et al.*, 1979).

Subgroups of Natriboralfs have not been developed in the United States Taxonomy. Most of those in areas close to the boundary between the boreal forest and grasslands probably should be classified as *Cryic Natriboralfs*. For soils underlain by permafrost, as in eastern Siberia, a subgroup of *Pergelic Natriboralfs* will be required.

B. AQUALFS

Aqualfs have characteristics of wetness beyond the mottling and concretions in the albic horizon and the upper part of the argillic horizon that define the Aquic Cryoboralfs. The argillic horizon in Aqualfs is saturated most of the summer and exhibits the gray or bluish colors that accompany prolonged wetness. Saturation of the argillic horizon may be caused by a consistently high groundwater table or may be a consequence of very slowly permeable parent materials that retain moisture from precipitation long after the surface horizons have lost excess water to transpiration and evaporation (Liverovskiy and Roslikova, 1962). In every case, however, there is a period of relative dryness in the solum that permits operation of the argilluviation process alternately with the gleying process (Pawluk, 1971). The soil below the solum is commonly saturated throughout the year.

Aqualfs occur in association with Cryoboralfs wherever they exist in the boreal belt. They may be in depressions or, in the cold, humid parts of the belt where evapotranspiration rates are low and the subsoil remains frozen until late summer, on level uplands with little runoff (Rybtsov, 1960). The vegetation may be dominantly coniferous forest (Rusanova *et al.*, 1975), mixed forest (Ufimtseva, 1970), or shrubs, sedges, and grasses (Liverovskiy and Roslikova, 1962). In most instances, mosses dominate the ground cover and the 0 horizon is peaty. Few, if any, Aqualfs exist in tundra areas because of shallow permafrost and cryoturbation processes that prevent the development of an argillic horizon.

In the warmer part of the boreal belt the argillic horizon in Aqualfs has stronger blocky structure and thicker clay skins than in associated Boralfs (Acton and St. Arnaud, 1963), but in areas adjacent to the arctic tundra, albic horizons are thinner, and clay films in the Aqualfs are not as well-developed (Archegova, 1974; Rusanova, 1976). Reducing conditions in the albic and argillic horizons keep iron in more soluble form, and clay skins, as a result, have bluish colors rather than the brown colors typical of clay skins in the Cryoboralfs. In the upper horizon of Aqualfs, a higher proportion of the iron is extractable by acid ammonium oxalate; the ratio between oxalate-extractable and dithionite-extractable iron in the Aqualfs is narrower than .35, whereas in the Boralfs it is wider than .35 (Stonehouse and St. Arnaud, 1971). This is a further indication that less of the total iron exists in crystallized form in the Aqualfs.

At present, no great group of Cryaqualfs is recognized in the United States Taxonomy, largely because they have not been observed in the United States. Later revisions of the taxonomy probably will correct this omission. Subgroups should include *Typic Cryaqualfs* for soils with characteristics like those of Typic Cryoboralfs except for the properties related to wetness; *Glossic Cryaqualfs* for soils with tonguing of the albic horizon (Rusanova, 1976); *Pergelic Cryaqualfs* for soils underlain by permafrost but at depths great enough that cryoturbation is minimal (Firsova, 1967); and, possibly, *Paleo Cryaqualfs* for soils of ''monsoon'' climates with very thick albic horizons (Liverovskiy and Roslikova, 1962). It is likely that subgroups of *Cryic Natraqualfs* and *Pergelic Natraqualfs* will also be required for poorly drained soils with natric horizons.

References

Acton, D. F., and St. Arnaud, R. J. (1963). Micropedology of the major profile types of the Weyburn catena. *Can. J. Soil Sci.* **43**, 377–386.

Anderson, D. W., DeLong, E., and McDonald, D. S. (1979). The pedogenetic origin and characteristics of organic matter of Solod soils. *Can. J. Soil Sci.* **59**, 357–362.

Archegova, I. B. (1974). Humus profiles of some taiga and tundra soils in the European U.S.S.R. *Sov. Soil Sci.* **6**, 136–141.

Aristovskaya, T. V. (1963). Decomposition of organic mineral compounds in podzolic soils. *Sov. Soil Sci.,* 20–29.

Bray, R. H. (1934). A chemical study of soil development in the Peorian loess region of Illinois. *Am. Soil Surv. Assn. Bull.* **15,** 58–65.

Buol, S. W., and Hole, F. D. (1961). Clay skin genesis in Wisconsin soils. *Soil Sci. Soc. Am. Proc.* **25,** 377–379.

Chernov, V. P. (1965). Typical podzolic soils of the Perm Oblast formed on clay loam moraine and mantle. *Sov. Soil Sci.,* 207–218.

Chichester, F. W., Youngberg, C. T., and Harward, M. E. (1969). Clay mineralogy of soils formed in Mazama pumice. *Soil Sci. Soc. Am. Proc.* **33,** 115–120.

Clark, J. S., and Green, A. J. (1964). Some characteristics of gray soils of low base saturation from northeastern British Columbia. *Can. J. Soil Sci.* **44,** 319–328.

Dobrovol'skiy, G. V., Afanas'yeva, T. V., Vasilenko, V. I., Devirts, A. L., and Markova, N. G. (1970). Genesis and age of secondary podzolic soils in West Siberia. *Sov. Soil Sci.* **2,** 291–293.

Duchafour, P. (1951). Lessivage et podzolization. *Rev. Forest. Franc., No. 10,*

Dudal, R. (1968). "Definition of Soil Units for the Soil Map of the World." Food and Agr. Org. (FAO) Pub. 33. Rome.

Filimonova, L. C. (1965). Characteristics of the taiga soils of the Aldan and Tommot Rayons of the Yakut A. S. S. R. *Sov. Soil. Sci.,* 219–225.

Firsova, V. P. (1967). Forest soils of the northern Transural region. *Sov. Soil Sci.,* 300–307.

Floate, M. J. S. (1966). A chemical, physical, and mineralogical study of soils developed on glacial lacustrine clays in north central British Columbia. *Can. J. Soil Sci.* **46,** 227–236.

Frei, E. (1967). Mikromorphologische und chemische Untersuchungen eines entbosten Parabraunerdeprofils des schweizerischer Mittellandes und Deutung seiner entwicklungstendenz. *Geoderma* **1,** 289–298.

Fridland, V. M. (1958). Podzolization and illimerization (clay migration). *Sov. Soil Sci.,* 24–32.

Gradusov, B. P., and Dyazdevich, G. S. (1961). Chemical and mineralogical composition of clay fractions in strongly podzolic soils in connection with element migration. *Sov. Soil Sci.,* 749–756.

Harlan, P. W., Franzmeier, D. P., and Roth, C. B. (1977). Soil formation on loess in southwestern Indiana: II. Distribution of clay and free oxides and fragipan formation. *Soil Sci. Soc. Am. Jour.* **41,** 99–103.

Jungerius, P. D. (1969). Soil evidence of postglacial tree line fluctuations in the Cypress Hills area, Alberta, Canada. *Arc. Alp. Res.* **1,** 235–246.

Karpachevskiy, L. O. (1960). Micromorphological study of leaching and podzolization of soils in a forest. *Sov. Soil Sci.,* 493–500.

Kremer, A. M. (1969). Microstructure of strongly podzolic soil and movement of clay suspensions. *Sov. Soil Sci.* **1,** 286–293.

Kuz'min, V. A. (1969). Organic matter of separates in the Sod-podzolic soils with a second humus horizon on the Oka-Angara interfluve. *Sov. Soil Sci.* **1,** 278–285.

Kuz'min, V. A. (1973). Mountain soils of the Cis-Baikal. *Sov. Soil Sci.* **5,** 513–523.

Leahey, A. (1947). Characteristics of soils adjacent to the Mackenzie River in the Northwest Territories of Canada. *Soil Sci. Soc. Am. Proc.* **12,** 458–461.

Liverovskiy, Yu. A., and Roslikova, V. I. (1962). Genesis of some meadow soils in the Maritime Territory. *Sov. Soil Sci.,* 814–823.

Lowe, L. E. (1980). Humus fraction ratios as a means of discriminating between horizon types. *Can. J. Soil Sci.* **60,** 219–229.

Makeyev, O. V., and Nogina, N. A. (1958). Genus of Sod-forest soils on residual-alluvial traps. *Sov. Soil Sci.,* 778–786.

McKeague, J. A., and Cann, D. B. (1969). Chemical and physical properties of some soils derived from reddish brown materials in the Atlantic provinces. *Can. J. Soil Sci.* **49,** 65–78.

McKeague, J. A., and St. Arnaud, R. J. (1969). Pedotranslocation: Eluviation–illuviation in soils during the Quaternary. *Soil Sci.* **107,** 428–434.

McKeague, J. A., Nowland, J. L., Brydon, J. E., and Miles, N. M. (1971). Characterization and classification of five soils from eastern Canada having prominently mottled B horizons. *Can J. Soil Sci.* **51**, 483–497.

McKeague, J. A., Miles, N. M., Peters, T. W., and Hoffman, D. W. (1972). A comparison of luvisolic soils from three regions in Canada. *Geoderma* **7**, 49–69.

McKeague, J. A., MacDougall, J. I., and Miles, N. M. (1973). Micromorphological, physical, chemical, and mineralogical properties of a catena of soils from Prince Edward Island in relation to their classification and genesis. *Can. J. Soil Sci.* **53**, 281–295.

Ovchinnikov, S. M., Sokolova, T. A., and Targul'yan, V. O. (1973). Clay minerals in loamy soils in the taiga and forest tundra of West Siberia. *Sov. Soil Sci.* **5**, 709–722.

Pawluk, S. (1961). Mineralogical composition of some Grey Wooded soils developed from glacial till. *Can. J. Soil Sci.* **41**, 228–240.

Pawluk, S. (1971). Characteristics of Fera Eluviated Gleysols developed from acid shales in north-western Alberta. *Can. J. Soil Sci.* **51**, 113–124.

Pawluk, S., and Dudas, M. (1978). Reorganization of soil materials in the genesis of an acid luvisolic soil of the Peace River region, Alberta. *Can. J. Soil Sci.* **58**, 209–220.

Pedro, G., Jamagne, M., and Begon, J. C. (1978). Two routes in genesis of strongly differentiated acid soils under humid, cool-temperate conditions. *Geoderma* **20**, 173–189.

Pettapiece, W. W. (1975). Soils of the subarctic in the lower Mackenzie basin. *Arctic* **28**, 35–53.

Pettapiece, W. W., and Pawluk, S. (1972). Clay mineralogy of soils developed partially from volcanic ash. *Soil Sci. Soc. Am. Proc.* **36**, 515–519.

Pettapiece, W. W., and Zwarich, M. A. (1970). Micropedological study of a Chernozemic to Grey Wooded sequence of soils in Manitoba. *J. Soil Sci.* **21**, 138–145.

Polteva, R. N., and Sokolova, T. A. (1967). Investigation of concretions in a strongly podzolic soil. *Sov. Soil Sci.,* 884–893.

Pringle, W. L., Cairns, R. R., Hennig, A. M. F., and Siemens, B. (1975). Salt status of some soils of the Slave River lowlands in Canada's Northwest Territories. *Can. J. Soil Sci.* **55**, 399–406.

Ranney, R. W., and Beatty, M. T. (1969). Clay translocation and albic tongue formation in two Glossoboralfs of west-central Wisconsin. *Soil Sci. Soc. Am. Proc.* **33**, 768–775.

Reeder, S. W., Arshad, M. A., and Odynsky, W. (1967). The relationship between structural stability and chemical criteria of some solonetzic soils of northwestern Alberta. *Can. J. Soil Sci.* **47**, 231–237.

Rozov, N. N., and Ivanova, Ye. N. (1967). Classification of the soils of the U. S. S. R. (Principles and a systematic list of soil groups). *Sov. Soil Sci.,* 147–155.

Rusanova, G. V. (1976). Distribution and chemistry of clay incrustations in podzolic soils in relation to their genesis. *Sov. Soil Sci.* **8**, 133–143.

Rusanova, G. V., Tsypanova, A. N., and Bushuyeva, Ye. N. (1975). Content and some properties of concretions in the podzolic soils of the central taiga subzone of the Komi A. S. S. R. *Sov. Soil Sci.* **7**, 272–280.

Rybtsov, D. M. (1960). Podzolic-Bog soils. *Sov. Soil Sci.,* 705–714.

Sawyer, C. D., and Pawluk, S. (1963). Characteristics of organic matter in degrading chernozemic surface soils. *Can. J. Soil Sci.* **43**, 275–286.

Soil Survey Staff. (1975). "Soil Taxonomy." U. S. Dept. Agr. Handbook 436, Washington, D.C.

Sokolov, I. A., Tursina, T. V., and Belousova, N. I. (1969). Present-day podzolization on the plains of central Yakutia. *Sov. Soil Sci.* **1**, 657–663.

Sokolov, I. A., Gradusov, B. P., Tursina, T. V., Tsyurupa, I. G., and Tyapkina, N. A. (1974). Description of soil formation on unconsolidated silicate rocks in the permafrost-taiga region. *Sov. Soil Sci.* **6**, 269–282.

Sokolov, I. A., Naumov, Ye. M., Gradusov, B. P., Tursina, T. V., and Tsyurupa, I. G. (1976). Ultracontinental taiga soil formation on calcareous loams in central Yakutia. *Sov. Soil Sci.* **8**, 144–160.

Stobbe, P. C. (1952). The morphology and genesis of the Gray-Brown Podzolic and related soils of eastern Canada. *Soil Sci. Soc. Am. Proc.* **16**, 81-84.

Stonehouse, H. B., and St. Arnaud, R. J. (1971). Distribution of iron, clay and extractable iron and aluminum in some Saskatchewan soils. *Can. J. Soil Sci.* **51**, 283-292.

Tolchel'nikov, Yu. S. (1970). Contribution to the description of the absolute age of the second horizon of Sod-Podzolic soils in West Siberia. *Sov. Soil Sci.* **2**, 289-290.

Ufimtseva, K. A. (1968). Present and relict properties of West Siberian lowland soils. *Sov. Soil Sci.,* 586-594.

Ufimtseva, K. A. (1970). Soils of the southern taiga subzone of the West Siberian plain. *Sov. Soil Sci.* **2**, 168-179.

Williams, B. H., and Bowser, W. E. (1952). Gray Wooded soils in parts of Alberta and Montana. *Soil Sci. Soc. Am. Proc.* **16**, 130-133.

Wright, J. R., Leahey, A., and Rice, H. M. (1959). Chemical, morphological and mineralogical characteristics of a chronosequence of soils on alluvial deposits in the Northwest Territories. *Can. J. Soil Sci.* **39**, 32-43.

Yelovskaya, L. G. (1965). Saline soils of Yakutia. *Sov. Soil Sci.,* 355-359.

Zaboyeva, I. V. (1958). Gley-podzolic soils. *Sov. Soil Sci.* 237-244.

Zaboyeva, I. V. (1965). Gley-podzolic soils of northeast European U.S.S.R. *Sov. Soil Sci.,* 745-755.

Zonn, S. V. (1966). Development of Brown Earths, Pseudopodzols, and Podzols. *Sov. Soil Sci.,* 751-758.

Zonn, S. V. (1973). Environmental settings of the processes of lessivage, pseudopodzolization, and podzolization during the Quaternary period in the western and northwestern regions of the U. S. S. R. *Soil Sci.* **116**, 211-217.

Chapter 7

MOLLISOLS

Mollisols are soils with mollic epipedons, but some soils with mollic epipedons may be excluded from the order of Mollisols because of the presence of other diagnostic horizons or characteristics. For this reason the definition of the order is fairly complicated. In cold regions, however, most soils with mollic epipedons are Mollisols. The principal exception is soils in which the mollic epipedon is formed in volcanic ash thicker than 35 cm; those soils are classified as Andepts. Soils with a mollic epipedon but with less than 50% base saturation below the epipedon or with a spodic horizon at a depth of less than 2 m are also excluded, but these situations are rare in cold regions.

Soils that have the requisite total thickness for a mollic epipedon but in which an albic horizon separates the upper from the lower part of the epipedon are considered to be Mollisols. A histic epipedon is permitted if it occurs above a mollic epipedon that has formed in mineral soil materials. Cambic, argillic, natric, or calcic subsurface horizons may occur below the mollic epipedon, or the epipedon may overlie unaltered soil material.

I. Genesis of Mollisols

A. CONDITIONS OF FORMATION

Almost invariably, the parent material of cold Mollisols is either calcareous or is derived from basalt or other basic rock. The soils occur in a wide variety of settings and support several kinds of vegetation. In the transition zone between the boreal forest and subhumid grasslands of the northern Great Plains of North America, they support either grassy (Dudas and Pawluk, 1969) or mixed forest (Dumanski and Pawluk, 1971) vegetation. In the Rocky Mountains they occur under grasses and willows (Retzer, 1956) or, on limestone, under a sparse alpine vegetation or a gravelly pavement with little higher vegetation (Nimlos and McConnell, 1965). In southern Chile they occupy semiarid grasslands in the rain shadow of the Andes (Mikhaylov, 1970). Under the cold continental climate of the southern Siberian mountains, Mollisols occur in grass-covered basins and valleys (Sokolov and Sokolova, 1962; Sokolov, 1967; Volkovintser, 1969); on lower portions of steep, south-facing slopes under grass or mixed grass and forest (Ufimtseva, 1964; Kuz'min and Chernegova, 1976); in forested areas in intermontane depressions (Dimo, 1965); and on some forested higher and north-facing slopes (Nadezhdin, 1958; Filimonova, 1965; Gerasimov, 1965). Farther north they occupy portions of grassy depressions in association with saline soils in the dry plains of eastern Siberia (Gerasimov, 1965) and grassy south-facing slopes in the northern mountains (Naumov and Andreyeva, 1963; Volkovintser, 1974). They are common on basic materials in the arctic tundra, under both well-drained and poorly drained conditions (Ignatenko, 1963; Rieger, 1966; Ignatenko, 1967; Mikhaylov, 1960; Everett and Parkinson, 1977; Rieger et al., 1979).

B. DEVELOPMENT OF THE MOLLIC EPIPEDON

The mollic epipedon is formed by accumulation, decomposition, and dispersal of organic matter in the upper part of the mineral soils. This can take place in several ways. In the warmer part of the boreal belt, the development sequence, known as the "sod process" in the Russian literature, is essentially the same as that in the temperate zone. Cations such as calcium and magnesium are taken up by plants with fairly deep rooting systems and eventually are deposited in organic litter at the soil surface and, to a lesser extent, in the upper mineral horizon as fine roots die. Decomposition of the organic matter and its incorporation into the upper soil horizons by faunal activity is rapid because of the high nutrient status of the soil, and the cations and other nutrients are soon available for recycling through the vegetation. The A horizon thus formed thickens and darkens until an equilibrium between accumulation and the amount lost to biological use of the organic matter is reached. Under stable environmental conditions the horizon, which may now be thick enough to constitute a mollic epipedon, retains its high base status and has a crumb or soft granular structure that permits moisture to penetrate readily but resists removal and leaching of clay particles. Calcium and other cations that are in excess of the exchange capacity of the soil may be transported in solution to layers below the root zone, but the production of soluble organic acids is minimal.

In somewhat colder areas the rate of decomposition is not as great and, because of shallower root systems, the rate of replenishment of cations at the surface is reduced. Although well-decomposed organic matter is incorporated into the upper mineral horizon as in warmer soils, more of the organic matter remains in the litter on the soil surface, and more organic acids are produced. In soils with moisture perched above a deep permafrost table, there is commonly an accumulation of iron in the A horizon, in part a result of weathering and biological activity and in part because of upward movement of dissolved ferrous iron during freezing (Makeyev and Nogina, 1958; Gerasimov, 1965). Leaching effectiveness is greater because of reduced evapotranspiration rates. Except in situations where fresh calcareous material is deposited regularly on the soil (Dumanski and Pawluk, 1971), the mollic epipedons are thinner and are more subject to degradation.

In very cold areas such as the tundra, where biological activity is low, mollic epipedons are slow to form even in basic materials but, once developed, persist for long periods of time. They exist, commonly, in association with Inceptisols or Entisols in more acidic parent materials (Retzer, 1956, 1974; Nimlos and McConnell, 1965; Rieger, 1966). Some Mollisols in tundra areas are subject to intense cryoturbation and disruption of the epipedon by subsoil material forced to the surface. In coastal plains and other low-lying areas, the mixed organic and mineral matter that makes up the epipedon may have accumulated in shallow lake basins and later become part of upland surfaces as a result of lake draining and other geomorphic processes (Everett and Parkinson, 1977). In mountains the epipedon may be a residue that remains after leaching of soluble components of limestone or shales that contain organic matter (Ugolini and Tedrow, 1963; Ugolini *et al.,* 1963).

C. DEGRADATION OF THE MOLLIC EPIPEDON

Continued leaching of well-drained Mollisols in forested areas or, at the warm limit of the boreal belt, a change in climate involving increased precipitation, cooler temperatures, and replacement of grassland by forest, results in acidification of the mollic epipedon. This leads, eventually, to degradation of the epipedon. A first step in this sequence is the translocation of free calcium carbonate to greater depths and the development of a brown or yellowish brown cambic horizon between the mollic epipedon and the underlying material. Soils in which this has taken place have been identified as "leached Chernozems" in the U.S.S.R. and as "degraded Chernozems" in western Europe (Dudas and Pawluk, 1969). A horizon of accumulation of gypsum may underlie the zone of maximum accumulation of calcium carbonate (Volkovintser, 1974). The color of the cambic horizon is due mainly to hydrated iron and aluminum oxides released in place by weathering of primary minerals (Makeyev and Nogina, 1958). As leaching, the loss of bases, and acidification continue, there is increasing dispersion of granules in the mollic epipedon and the onset of clay translocation, or argilluviation (Nadezhdin, 1958; Sawyer and Pawluk, 1963). The migrating clay consists mostly of illite and smectites but may include crystalline (dithionite-extractable) iron oxides (Stonehouse and St. Arnaud, 1971) (Fig. 25). In time the organic matter content of the surface horizon may be reduced sufficiently so that a mollic epipedon can no longer be recognized. Where climatic change and resultant argilluviation have resulted in the formation of an argillic horizon whose upper boundary is within the original mollic epipedon and in the creation of an albic horizon by water periodically perched above it, the mollic epipedon may be split into two parts—one at the surface and one below the albic horizon and coincident with the upper part of the argillic horizon. In coarse-textured soils subject to increased leaching, acidification can progress to the point that the spodic process can operate, but few cold soils that originally had mollic epipedons reach this stage of degradation.

The direction of development of Mollisols at the warm limit of the boreal belt following climatic change is illustrated by the soils that border them. Mollisols in dry valleys and the lower portions of south-facing slopes in the mountains of southern Siberia grade into Cryochrepts and Cryoboralfs at higher elevations and on northerly slopes (Sokolov and Sokolova, 1962; Ufimtseva, 1963; Kuz'min, 1973). In the Canadian Great Plains, Cryoboralfs occur just north of the Mollisols (Sawyer and Pawluk, 1963). In an area of continuing loess deposition east of the Rocky Mountains, Mollisols give way to first Cryochrepts and then Cryoboralfs with increasing distance from the source of the loess (Dumanski and Pawluk, 1971).

D. ORGANIC ACIDS

Typically, in temperate climates, the soluble fraction of organic matter in Mollisols is dominated by humic acid. This is true, also, in most Mollisols on the warm fringe of the boreal belt, but the proportion of fulvic acid in the mollic epipedon is higher, and fulvic acid may be dominant in the relatively small amount of organic matter in deeper horizons (Kuz'min and Chernegova, 1976). In colder soils fulvic acid may be the dominant soluble organic component in all horizons (Assing, 1960; Mikhaylov, 1960; Sawyer and Pawluk, 1963; Ignatenko, 1967; Volkovintser, 1969). Where humic acid occurs in cold, dry soils, its structure is less complex, and it is more mobile than humic acid in temperate soils (Volkovintser, 1974).

E. CRYOTURBATION

Frost action is commonly severe in well-drained Mollisols with permafrost. Where the permafrost table is deep, fissures extending from the surface may fill

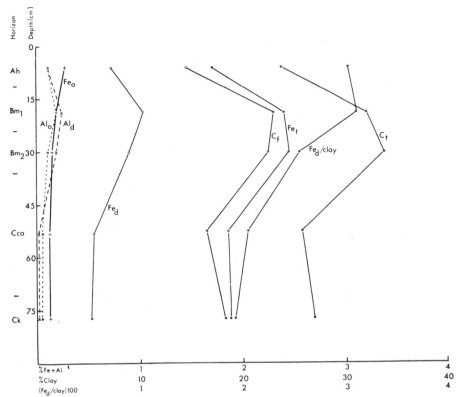

FIGURE 25. *Distribution of total iron (Fe$_t$), oxalate-extractable iron (Fe$_o$) and aluminum (Al$_o$), dithionite-extractable iron (Fe$_d$) and aluminum (Al$_d$), total < .002 mm clay (C$_t$), and fine (< .0002 mm) clay (C$_f$) in a Mollisol from Saskatchewan. (Reprinted from Stonehouse and St. Arnaud [1977] by permission of the Agricultural Institute of Canada.)*

with A-horizon material (Yelovskaya, 1965). In this case, the organic matter at depth may have as high a proportion of humic acid as the surface horizon (Dimo, 1965). In tundra areas cryoturbation may create large barren patches in which the mollic epipedon is disrupted (Ignatenko, 1967).

F. DEVELOPMENT OF MOLLISOLS UNDER POORLY DRAINED CONDITIONS

The mollic epipedon in poorly drained Mollisols may consist of material containing organic matter that has washed in from surrounding uplands as well as humus-rich material formed in place. The parent materials, although always high in bases, may or may not be calcareous or saline. Most poorly drained Mollisols have strongly mottled cambic horizons with no evidence of clay accumulation (Mikhaylov, 1960; Sokolov *et al.*, 1973), but some have horizons of accumulation of calcium carbonate, sodium salts, or eluviated iron at depth. A few have albic and argillic horizons below the mollic epipedon (Sokolov *et al.*, 1973). Some Mollisols of cold continental areas with low precipitation are only intermittently wet as a result of a fluctuating water table.

Where permafrost is shallow, poorly drained Mollisols are almost continuously wet in the summer. A histic epipedon commonly overlies the mollic epipedon. The mollic epipedon itself may be calcareous (Day and Rice, 1964). Frost action may result in deep tongues of material from both the histic and mollic epipedons and an accumulation of organic matter at the permafrost table (Sokolov and Sokolova, 1962), and in disruption of the epipedons by intrusions of gleyed subsurface material (Everett and Parkinson, 1977). Because of the instability of the soils, no argillic horizons can form.

II. Classification of Cold Mollisols

The two principal great groups of Mollisols that occur in cold regions are the Cryoborolls and Cryaquolls. The suborders of Albolls and Rendolls are also represented but are of minor extent. Many—perhaps too many—subgroups have been defined; most, but not all, are discussed below.

A. BOROLLS

Most well-drained cold Mollisols are Cryoborolls. *Typic Cryoborolls* are soils that have a mollic epipedon but no subsurface diagnostic horizons other than a brown cambic horizon and calcic and gypsic horizons deep in the profile. Also, soils are excluded from the Typic subgroup if the mollic epipedon is thicker than 40 cm unless it is formed in sandy material, the mean annual soil temperature is 0° C or colder, a layer of volcanic material thicker than 18 cm occurs in the upper part of the soil, or the depth to bedrock is shallower than 50 cm. Most Typic Cryoborolls have loamy textures. They occur primarily in areas near the transition between semiarid or subhumid grasslands and the boreal forest (Dudas and Pawluk, 1969) and on grass and shrub-covered, steep, south-facing slopes in generally forested areas (Volkovintser, 1974).

Calcic Cryoborolls have accumulations of calcium carbonate in sufficient quantity to constitute a calcic horizon either in the lower part of the mollic epipedon or immediately below the epipedon. They occur mostly in dry, cold continental areas and support grass or shrub vegetation (Nadezhdin, 1958; Sokolov, 1967; Yarilova *et al.*, 1970) but also occur under forest in areas with greater precipitation where the parent material is highly permeable weathered limestone (Filimonova, 1965; Gerasimov, 1965). Thick, calcareous mollic epipedons also form where there is continuing deposition of loess or other sediments and simultaneous accumulation of organic matter (Dumanski and Pawluk, 1971). If the mollic epipedon is thicker than 40 cm (except in sandy materials), the soil is assigned to a separate subgroup of *Calcic Pachic Cryoborolls.*

Andeptic Cryoborolls have a layer of volcanic ash or tuff 18–35 cm thick in the upper part of the soil. Some of the volcanic material may have weathered to amorphous clays. These soils differ from Cryandepts of relatively dry climates only in that the volcanic deposit is thinner.

Cold Mollisols in which argillic horizons have developed are distributed among several rather finely defined subgroups. Soils with an argillic horizon immediately below the mollic epipedon and with no albic horizon are classified as *Argic Cryoborolls.* If leaching has resulted in the formation of an albic horizon between the epipedon and the argillic horizon, the soils are *Boralfic Cryoborolls.* If there is an increase in clay of at least 20% in the upper part of the argillic horizon as compared with the overlying horizon and the albic horizon has formed in part as a result of the intermittent presence of perched water above the argillic horizon, they are *Abruptic Cryoborolls.* Soils in which a natric horizon occurs in place of the argillic horizon, with or without an albic horizon, are classified as *Natric Cryoborolls.* These soils commonly occur in depressions in areas dominated by either Boralfic Cryoborolls or Cryoboralfs.

Vertic Cryoborolls are fine-textured Mollisols of cold continental areas with marked wet and dry seasons that have a network of fissures on the surface, deep humus tongues, and streaks of humus throughout the soil. Although fine-textured, they have no argillic horizon. Vertic Cryoborolls have many properties in common with soils of the order of Vertisols (all of which have temperature regimes warmer than cryic) but lack the smooth fracture planes, or slickensides, of those soils. Unlike the Vertisols, in which soil movements result primarily from shrinking and swelling of fine clay because of changes in soil moisture, the Vertic Cryoborolls owe their irregular distribution of organic matter in large part to frost action.

Pergelic Cryoborolls have mean annual temperatures of 0° C or less and, therefore, permafrost at some depth. They occur under grass–shrub, forest, and tundra vegetation. Under both grass–shrub cover in the driest parts of the boreal belt (Volkovintser, 1974) and mixed forest adjacent to the grasslands (Sokolov and Sokolova, 1962; Dimo, 1965; Kuz'min, 1973), the permafrost table is generally 2 m or more deep, and humus tongues extending to the permafrost are common. Under alpine meadow vegetation in mountains at lower latitudes, the permafrost table is also deep, but the mollic epipedons are not as thick, and tonguing is less pronounced (Retzer, 1974). In tundra areas, both in the arctic and in high mountains, most of the Pergelic Cryoborolls are subject to intense frost action (Igantenko, 1963, 1967; Rieger, 1966). Where the mollic epipedon exists only under vegetation and is absent in barren boils or frost circles, the soils are classified as *Pergelic Ruptic-Entic Cryoborolls.* (This subgroup, although not yet given official status in the United States Taxonomy, is fairly extensive.) On some high ridges, especially on weathered limestone, the mollic epidedon may be continuous although higher vegetation covers only a small portion of a dry rubbly surface (Nimlos and McConnell, 1965). The organic matter in the epipedon in this case may be derived from lichens and microflora or may be a residue after leaching of soluble components of the limestone (Ugolini and Tedrow, 1963).

Lithic Cryoborolls have bedrock within 50 cm of the surface. Mean soil temperatures may be higher or lower than 0° C, and vegetation may be forest or tundra (Targul'yan, 1959). Over limestone, especially, an argillic horizon may occur intermittently above the bedrock; such soils are in a subgroup of *Lithic Ruptic–Argic Cryoborolls*. If, because of frost action or other cause, the mollic epipedon is not continuous within each pedon, the soils are classified as *Lithic Ruptic–Entic Cryoborolls*.

Aquic Cryoborolls are mottled at some depth below the mollic epipedon because of an intermittently high water table but are not wet enough to have developed the characteristics of Cryaquolls. They occur mostly in floodplains. Soils in which water is perched temporarily over deep permafrost may also be mottled at depth (Sokolov and Sokolova, 1962; Sokolov, 1967), but such soils are included with the Pergelic Cryoborolls.

B. ALBOLLS

Soils in which perched water above a slowly permeable argillic horizon lingers long enough to produce mottles or iron–manganese concretions in an albic horizon are separated in the suborder of *Albolls*. The albic horizon in Albolls may be entirely below the mollic epipedon or may separate upper and lower portions of the epipedon; the lower portion in this case coincides with the upper part of the argillic horizon. No *Cryalbolls* are recognized in the United States Taxonomy at the present time, but it is likely that some exist, especially where parent materials are saline.

C. RENDOLLS

Rendolls are soils in which a mollic epipedon has formed directly above highly calcareous, unconsolidated materials. In cold regions they form in glacial till derived from limestone, in colluvial detritus from limestone outcrops (Tedrow and Walton, 1977), and over recently elevated accumulations of seashells (Payette and Morisset, 1974). Organic matter accumulations in some cases may be high enough to approach the requirements for Folists (Chapter 12). No suborders have been defined in the United States Taxonomy. Cold

Rendolls, with or without permafrost, are classified as *Cryic Rendolls* or, if bedrock is shallow, as *Cryic Lithic Rendolls.*

Cryaquolls are the poorly drained Mollisols of cold regions. *Typic Cryaquolls* are those with no diagnostic horizons other than the mollic epipedon and a mottled cambic horizon and with no permafrost. They occur principally in depressions in the warmer part of the boreal belt, in association with Typic Cryoborolls. Still wetter soils, in which a histic epipedon has formed above the mollic epipedon are *Histic Cryaquolls.* Where a calcic horizon occurs within or immediately below the mollic epipedon, the soils are separated as *Calcic Cryaquolls.* A mottled argillic or natric horizon below the mollic epipedon, with or without a deeper calcic horizon, defines the subgroup of *Argic Cryaquolls.* All but the Typic subgroup are inextensive.

Pergelic Cryaquolls occur commonly under coniferous or grass–sedge vegetation in continental areas and under alpine or arctic tundra. The forested and grass-covered Pergelic Cryaquolls occupy depressions in which inflowing moisture is perched above a fairly deep permafrost table. Some are associated with salt-encrusted soils in the lowest parts of the depressions (Yelovskaya, 1965) and others with Histosols. In tundra areas they are continually wet because of moisture perched above a shallow permafrost table and commonly have a histic epipedon. Frost action in clayey materials may result in disruption of the mollic and histic epipedons and in small barren spots on the surface; a subgroup of *Pergelic Ruptic–Aqueptic Cryaquolls* is needed for classification of these complex soils (Everett and Parkinson, 1977) (Fig. 26).

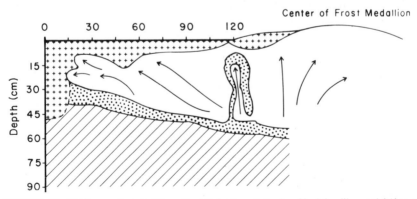

FIGURE 26. *Thickness and extent of the mollic epipedon (cross pattern) and buried mollic material (dot pattern) in a Pergelic Ruptic–Aqueptic Cryaquoll of arctic Alaska. The diagonal pattern represents perennially frozen material. Arrows indicate the probable direction of movement of material within the polygon in response to frost pressures. (From Everett and Parkinson [1977] by permission of University of Colorado.)*

References

Assing, I. A. (1960). Some features of humus formation in mountain soils of northern Tyan'-Shan.'. *Sov. Soil Sci.*, 1277–1282.

Day, J. H., and Rice, H. M. (1964). The characteristics of some permafrost soils in the Mackenzie Valley, N. W. T. *Arctic* **17**, 223–236.

Dimo, V. N. (1965). Formation of a humic-illuvial horizon on permafrost. *Sov. Soil Sci.*, 1013–1021.

Dudas, M. J., and Pawluk, S. (1969). Chernozem soils of the Alberta parklands. *Geoderma* **3**, 19–36.

Dumanski, J., and Pawluk, S. (1971). Unique soils of the Foothills region, Hinton, Alberta. *Can. J. Soil Sci.* **51**, 351–362.

Everett, K. R., and Parkinson, R. J. (1977). Soil and landform associations, Prudhoe Bay area, Alaska. *Arc. Alp. Res.* **9**, 1–19.

Filimonova, L. G. (1965). Characteristics of the taiga soils of the Aldan and Tommot Rayons of the Yakut A. S. S. R. *Sov. Soil Sci.*, 219–225.

Gerasimov, I. P. (1965). Humus-calcareous cryogenic soils of the Aldan-Lena interfluve. *Sov. Soil Sci.*, 140–148.

Ignatenko, I. V. (1963). Arctic tundra soils of the Yugor Peninsula. *Sov. Soil Sci.*, 429–440.

Ignatenko, I. V. (1967). Soil complexes of Vaygach Island. *Sov. Soil Sci.*, 1216–1229.

Kuz'min, V. A. (1973). Mountain soils of the Cis-Baikal. *Sov. Soil Sci.* **5**, 513–523.

Kuz'min, V. A., and Chernegova, L. G. (1976). Brown Mountain-Taiga and Sod Forest soils in the Baykal region. *Sov. Soil Sci.* **8**, 373–380.

Makeyev, O. V., and Nogina, N. A. (1958). Genus of Sod Forest soils on residual-alluvial traps. *Sov. Soil Sci.*, 778–786.

Mikhaylov, I. S. (1960). Some features of Sod Arctic soils on Bolshevik Island. *Sov. Soil Sci.*, 649–657.

Mikhaylov, I. S. (1970). Main features of Chilean soils. *Sov. Soil Sci.* **2**, 1–7.

Nadezhdin, B. V. (1958). Sod-carbonate soils of the southern central Siberian plateau. *Sov. Soil Sci.*, 642–647.

Naumov, Ye. M., and Andreyeva, A. A. (1963). Soils of the Yana-Indigir upland slopes under steppe vegetation. *Sov. Soil Sci.*, 249–255.

Nimlos, T. J., and McConnell, R. C. (1965). Alpine soils in Montana. *Soil Sci.* **99**, 310–321.

Payette, S., and Morisset, P. (1974). The soils of Sleeper Islands, Hudson Bay, N. W. T., Canada. *Soil Sci.* **117**, 352–368.

Retzer, J. L. (1956). Alpine soils of the Rocky Mountains. *J. Soil Sci.* **7**, 22–32.

Retzer, J. L. (1974). Alpine soils. *In* "Arctic and Alpine Environments." (J. D. Ives and R. G. Barry, eds.), pp. 771–802. Methuen, London.

Rieger, S. (1966). Dark well-drained soils of tundra regions in western Alaska. *J. Soil Sci.* **17**, 264–273.

Rieger, S., Schoephorster, D. B., and Furbush, C. E. (1979). "Exploratory Soil Survey of Alaska." U. S. Dept. Agr., Soil Cons. Serv. Washington, D. C.

Sawyer, C. D., and Pawluk, S. (1963). Characteristics of organic matter in degrading chernozemic surface soils. *Can. J. Soil Sci.* **43**, 275–286.

Sokolov, I. A. (1967). Soils of the Transbaikal steppe. *Sov. Soil Sci.*, 319–330.

Sokolov, I. A., and Sokolova, T. A. (1962). Zonal soil groups in permafrost regions. *Sov. Soil Sci.*, 1130–1136.

Sokolov, I. A., Turchina, T. V., Smirnova, G. Ya., and Tyapkina, N. A. (1973). The origin of the bleached surface horizon in the permafrost soils of central Yakutia. *Sov. Soil Sci.* **5**, 287–299.

Stonehouse, H. B., and St. Arnaud, R. J. (1971). Distribution of iron, clay and extractable iron and aluminum in some Saskatchewan soils. *Can. J. Soil Sci.* **51,** 283-292.

Targul'yan, V. O. (1959). The first stages of weathering and soil formation on igneous rocks in the tundra and taiga zones. *Sov. Soil Sci.,* 1287-1296.

Tedrow, J. C. F., and Walton, G. F. (1977). Rendzina formation on Bathurst Island. *J. Soil Sci.* **28,** 519-525.

Ufimtseva, K. A. (1963). Mountain taiga soils of the Transbaikal region. *Sov. Soil Sci.,* 241-248.

Ufimtseva, K. A. (1964). Soils of the continental forest–steppe of southern Siberia. *Sov. Soil Sci.,* 783-789.

Ugolini, F. C., and Tedrow, J. C. F. (1963). Soils of the Brooks Range, Alaska: 3. Rendzina of the Arctic. *Soil Sci.* **96,** 121-127.

Ugolini, F. C., Tedrow, J. C. F., and Grant, C. L. (1963). Soils of the northern Brooks Range, Alaska: 2. Soils derived from black shale. *Soil Sci.* **95,** 115-123.

Volkovintser, V. I. (1969). Soil formation in the steppe basins of southern Siberia. *Sov. Soil Sci.* **1,** 383-391.

Volkovintser, V. I. (1974). Soils of the dry steppe of the Yana-Oymyakon upland. *Sov. Soil Sci.* **6,** 142-150.

Yarilova, Ye. A., Makeyev, O. V., and Tsybzhitov, Ts. Kh. (1970). Genetic and micromorphological characteristics of some soils in the western Baikal region. *Sov. Soil Sci.* **2,** 207-219.

Yelovskaya, L. G. (1965). Saline soils of Yakutia. *Sov. Soil Sci.,* 355-359.

Chapter 8

INCEPTISOLS: CRYANDEPTS

Andepts are soils developed in volcanic ash or, under some conditions, in weathered basaltic rock or other parent material that has a high vitric (glass) content and low bulk density. The vitric material must be at least 35 cm thick unless it rests directly on bedrock or other material that cannot be penetrated by roots. The soils are well- or moderately well-drained and have a mollic or umbric epipedon. Ashy soils with an aquic moisture regime are classified as Aquepts rather than Andepts, and recently deposited ash in which there has been no development of diagnostic horizons are considered to be Entisols. If the ash deposit is less than 35 cm thick over unconsolidated material, the classification of the soil is based on the properties of the buried material rather than on those of the vitric overburden.

I. Genesis of Cryandepts

A. CONDITIONS OF FORMATION

Andepts exist in all areas of the world that are affected by volcanic activity. Cryandepts occur principally in areas bordering the Bering Sea, in islands of the North Atlantic, and in high volcanic mountains, especially those along the west coast of North and South America. Apart from characteristics directly related to temperature, Cryandepts have essentially the same properties as the warmer Andepts (Simonson and Rieger, 1967). Most Cryandepts have black or dark reddish brown surface horizons and brown or yellowish brown subsoil horizons. They commonly occur in areas that have been subject to repeated ash falls. As a result buried sequa of surface and subsurface horizons and textural stratification are common. In general the coarser Cryandepts, including soils derived dominantly from cinders, occur close to the volcanoes and finer Cryandepts are at more distant sites, but strong winds blowing at the times of eruption can distort this pattern.

Most Cryandepts are in areas close to oceans and have developed under humid climates. Ash falls do occur in inland areas with dry continental climates, such as east central Alaska and Yukon Territory, but in most cases the total thickness of the ash in those areas is less than 35 cm, and development in the volcanic material has been minimal.

Cryandepts support several different kinds of vegetation. Under strongly maritime climates such as that of the Aleutian Islands and Iceland, the vegetation on low hills is dominated by grasses, alder, and associated forbs and shrubs, with mosses and lichens in the ground cover. Willows, cottonwood, and birch occupy some valley bottoms. At higher elevations and in level areas that experience frequent strong cold winds, low shrubs such as crowberry along with lichens and mosses are dominant (Ulrich, 1946; Rieger and Wunderlich, 1960; Helgason, 1963; Everett, 1977). Soils with this tundra-like vegetation, although lacking mottles and other characteristics of wetness, usually have a much higher moisture content than other Cryandepts and have a strongly mounded surface relief. In some areas with cold humid climates, such as Kamchatka, Cryandepts support a sparse birch forest with a grassy or shrubby undergrowth (Sokolov and Belousova, 1964; Sokolov and Karayeva, 1965). Cryandepts of high elevations at lower latitudes are covered by dwarf forests, tussock grasses, or, at the highest points, alpine tundra (Frei, 1978). In a few places they occur under coniferous forest that has recently advanced into previously existing grassland (Rieger and Wunderlich, 1960).

Major ash falls destroy existing above-ground portions of grasses and shrubs, but in a humid climate any sulfates or chlorides adsorbed by the ash during the eruption are soon leached out of the soil, and the vegetation quickly reestablishes itself (Liverovskiy, 1971). Following a large ash fall in northern Kodiak Island, Alaska, in June 1912, for example, the area became almost completely barren, but in the following year plant growth, except at the mouths of streams where the ash accumulation was more than 1 m thick, was as luxuriant as ever (Rieger and Wunderlich, 1960).

B. INITIAL WEATHERING OF ASH

Fresh ash has a very high specific surface, largely because of fine tubes and vesicles within the grains (Sokolov and Belousova, 1966; Espinoza et al., 1975;

Parfitt *et al.,* 1977). Although proportions vary, most ash is high in silicon and aluminum, relative to iron (Rieger, 1974; Singer and Ugolini, 1974; Wada and Harward, 1974), and is readily weatherable. Released aluminum, mainly in the form of hydroxides, has a strong affinity for humus that has not yet decomposed to form organic acids and combines with it in such a way that the organic matter is not open to enzyme attack and is stabilized against further degradation and leaching (Broadbent *et al.,* 1964; Wada and Harward, 1974). The aluminum bound with the organic matter is incapable of combination and coprecipitation with silica, thus limiting the possibility of formation of aluminosilicates (allophanes) and favoring the formation of opaline silica deposits in the surface horizon (Wada and Harward, 1974; Higashi and Wada, 1977). Because of the low rate of allophane formation, the surface horizon of ashy soils is seldom thixotropic (Sokolova and Belousova, 1966). Released iron has a lesser affinity for fresh organic matter and may form a ferruginous film on the outer surface of the ash grains or may be incorporated into an allophane-like amorphous compound (hisingerite) even in the presence of humus (Wada and Higashi, 1976). The bound humus held on the weathered surfaces of the ash grains accumulates rapidly and, in fine-grained ash, may eventually comprise 30% or more of the dry weight of the surface horizon (Ulrich, 1946; Wada and Higashi, 1976). Cations released by weathering neutralize organic acids formed in the litter and generally keep the pH above 5 even in humid climates (Sokolov and Belousova, 1964).

C. ALLOPHANE AND THIXOTROPY

In horizons that no longer have additions of fresh organic matter, such as those buried by more recent ash falls, and in deeper layers of thick single deposits, silica and aluminum are free to combine to form allophanes. This process is a central feature in the development of Andepts and takes place in all ashy soils in humid and subhumid environments. Allophanes are members of a series of hydrous aluminosilicates of varying composition that have in common a lack of repetitive crystallinity, a very high attraction for water, and, unlike crystalline layer silicates, a positive charge (Wada and Harward, 1974). The moisture-holding capacity of soils high in allophane may exceed 100% of the dry weight of the soil (Sokolov and Belousova, 1966; Everett, 1977). The moisture is held as hydrogen-bonded clusters of molecules around each particle; mechanical working of the soil breaks these clusters and leaves only single molecules of water on the grain surfaces (Wells and Furkert, 1972), but the clusters reform when the mechanical disturbance ceases. These transformations are responsible for the characteristic thixotropic nature of the subsurface horizons of fine-grained Andepts. Coarse or cindery ash deposits with little or no allophane are not thixotropic.

In older horizons of well-drained Andepts, soluble forms of silicon are lost by leaching, and some of the allophane is converted to imogolite (named for the brownish yellow "imogo" soils of Japan), a hydrous aluminum silicate with a weak layer crystalline structure (Wada and Harward, 1974). Under the electron microscope imogolite is seen to occur in threadlike formations, whereas allophane particles are spheroidal (Besoain, 1968). Imogolite occurs mostly in interstices between grains, and allophane principally within the grains (Wada and Kubo, 1975). The proportion and degree of development of imogolite is directly related to the age and depth of the horizons in which it occurs. In some old horizons it is dominant over allophane (Besoain, 1968).

The silica to alumina ratio in imogolite is about 1, in contrast with a ratio of about 1.7 in most allophanes (Wada and Kubo, 1975). Because its specific surface and moisture-holding capacity is much lower than that of allophane, older layers of ash deposits are less thixotropic than layers of intermediate age. In time, as desilication proceeds, at least a portion of the imogolite may be converted to crystalline hydrous aluminum oxide (gibbsite), reducing the thixotropy of the horizon still further. Under conditions of a limited leaching regime in which the silica concentration in the soil solution remains high, resilication and increased crystallization may convert the imogolite to halloysite and, eventually, kaolinite. This is a very slow process that occurs only in old, buried, ash horizons (Liverovskiy, 1971; Wada and Harward, 1974; McIntosh, 1980). Where horizons are alternately wet and dry, as in soils under climates with pronounced wet and dry seasons, metallic cations such as calcium, magnesium, and iron may combine with aluminum silicates to form montmorillonite (Chichester *et al.*, 1969). In humid climates where most Cryandepts are located, however, the persistently high moisture content of the soils and the abundant supply of fresh organic matter inhibit the formation of crystalline clay minerals (Sokolov and Belousova, 1964). Most fine-grained Cryandepts remain high in allophane and are strongly thixotropic.

E. DEGRADATION OF ANDEPTS

A final stage in the development of Cryandepts occurs in areas with cool, humid climates but with summers in which temperatures are high enough to permit substantial biological and chemical activity in the soil. Because rooting is shallow in such areas, there is little return of cations to the surface, and the organic litter and A horizon become increasingly acidic. Fulvic acid produced under these conditions combines with aluminum and iron to form organometallic complexes, and the spodic process comes into operation. This may take place under grassy or grass–alder vegetation (Rieger *et al.*, 1979) but occurs most commonly under coniferous forest (Liverovskiy, 1971). Degradation of Cryandepts to Spodosols is rapid following invasion of grasslands by conifers (Simonson and Rieger, 1967; Singer and Ugolini, 1974). The process is retarded, however, in areas that receive frequent deposits of volcanic ash (Liverovskiy, 1971). In many cases, because of the low iron content of the volcanic ash and the large supply of organic matter in humid areas, Spodosols formed in ashy materials are Humods (Rieger, 1974).

The existence of an albic horizon is the clearest indication that a soil has completed the transition to a Spodosol—although care must be exercised to be sure that a light-colored, recent ash deposit is not incorrectly assumed to be an albic horizon (Karpachevskiy, 1965)—but even where the characteristic horizon sequence of Spodosols has not developed many Andepts of cold, humid areas have indications of illuviation of humus, aluminum, and iron. Evidence for an intense humic–illuvial process, with accumulation of aluminum fulvates and amorphous iron oxides in the illuvial horizon, exists in ashy soils of Kamchatka that lack an albic horizon (Sokolov and Belousova, 1966). Some well-drained volcanic soils in the Aleutian Islands have concentrations of iron oxide below the highly organic surface horizon (Ulrich, 1946; Everett, 1977); soils with the morphology of Andepts in Kodiak Island, Alaska, have subsoils that satisfy the chemical requirements for a spodic horizon; and highly organic illuvial cutans coat peds in the lower part of thick umbric epipedons in ashy soils of high East African mountains (Frei, 1978). Because Andepts and well-developed Spodosols in ashy materials have many characteristics in common, positive identification may be difficult, but it seems likely that most Cryandepts of humid climates that do not receive additional ash deposits will eventually degrade to Spodosols. Where there are alternate wet and dry seasons and less effective leaching, the ultimate process may be argilluviation, and the Andepts may eventually be converted to Andeptic Cryoboralfs.

F. PROPERTIES OF ANDEPTS

The physical and chemical properties of Cryandepts, as in all Andepts, are affected by the dominant presence of allophanes in the clay fraction and the high proportion of organic matter in the soils. Their high porosity is reflected in low bulk density, which is always less than 1 g/cm^3, and in rapid percolation rates. The bulk density is lower in older than in recent ash because weathering and the removal of weathering products by leaching increases the porosity of the soil but does not decrease its volume. Bulk densities in weathered subsoil horizons are commonly in the range of .3 (Everett, 1977) to .5 g/cm^3 (Sokolov and Belousova, 1966).

The soils have a high pH-dependent cation exchange capacity and low base saturation. Despite this, pH values are generally above 5, except in some highly organic surface horizons. This combination of low base saturation and relatively high pH is a distinctive characteristic of Andepts. It may be due primarily to the high allophane content of the soils (Helgason, 1968) but may also be a function of their amorphous aluminum and iron hydroxide constituents (Wada and Harward, 1974).

Exchangeable calcium in Andepts is low in relation to magnesium because calcium is readily leached but magnesium from less weatherable ferromagnesian minerals that commonly are associated with volcanic ash and basalts is retained longer (Helgason, 1968). The soils have a very high phosphate-fixing ability, in part because of the large anion exchange capacity of allophane (Wada and Harward, 1974), and in part because of precipitation of insoluble iron and aluminum phosphates (Karpachevskiy, 1965).

In relatively young Andepts most plant nutrients are concentrated in the upper horizons. In older soils in which allophane has been partially replaced by halloysite, cations accumulate at the surface of the halloysite crystals. This results in reduced acidity of the soil solution and greater availability of nutrients to plants, and, as a consequence, deeper rooting and thickening of the A horizon (Liverovskiy, 1971).

Surface horizons are well-granulated because of the interaction between aluminum hydroxides and organic matter (Higashi and Waga, 1977). The horizon immediately below the highly organic surface horizon (the cambic or B horizon) in many Cryandepts develops a blocky structure. Deeper horizons, like those of other cold soils, have a platy structure (Helgason, 1963; Everett, 1977). The structure is easily destroyed by rubbing or shaking.

Andepts that are partially dried return to their original thixotropic condition on rewetting but, because of the crystallization of hydrous oxides, those taken to the air-dry condition lose their water-retaining capacity and become irreversibly aggregated (Sokolov and Belousova, 1966; Wells and Furkert, 1972; Wada and Harward, 1974). This situation is seldom encountered, however, in Andepts of cold areas.

II. Classification of Cryandepts

Classification of Cryandepts at the subgroup level in the United States Taxonomy is based largely on their thixotropic and water-retention characteristics. These depend principally on the allophane content of the soils, which in turn is related to the texture of the parent ash. Other factors, such as the amount of organic matter in the surface horizon and depth to bedrock, also are considered but affect relatively few soils. No Cryandepts with permafrost have been described as yet; if they exist, a Pergelic subgroup would be needed. The present classification is tentative and, because of its reliance on estimates of thixotropy, not entirely subject to quantitative verification. A proposal that Andepts, because of their unique properties, should be separated from the Inceptisols and reclassified as a new order of Andisols, is now under review by an international committee. It seems likely that the present classification of these soils in the United States Taxonomy will be subject to many modifications and perhaps to basic changes.

Typic Cryandepts in the existing taxonomy are well-drained soils in which 50% or more of the soil between depths of 25 and 100 cm (or between 25 cm and bedrock or other consolidated material at depths between 50 and 100 cm) consists of volcanic ash that has no thixotropic properties and a weighted average water retention against a tension of 15 bars that is less than 20%. Several different kinds of soils are included in the subgroup. Many consist completely of ash of fine sand-size or coarser that has been deposited fairly recently but is old enough that an umbric or mollic epipedon has formed at the surface. In some cases this horizon has been buried beneath a more recent and still relatively unweathered deposit less than 50 cm thick. A soil is also classified as a Typic Cryandept if it consists of stratified coarse ash and thixotropic finer ash but in which the nonthixotropic coarser material with low water-retention capacity makes up one-half or more of the thickness between the specified depth limits. Also, a soil is included in the subgroup if it is thixotropic to depths as great as 62 cm, but within that depth overlies nonvolcanic and nonthixotropic unconsolidated material with low water-retaining ability. Most properties of such a soil, especially as they relate to plant growth, are much more like those of the Dystric Cryandepts, described below, than of the Typic Cryandepts.

Typic Cryandepts generally occur fairly close to the volcanic source, where coarser ash particles are deposited. With increasing distance from the volcanoes, they grade into the finer Dystric Cryandepts. Because of differences in the particle-size distribution of erupted ash and the vagaries of wind speed and direction, however, the distance from the volcanoes at which this transition takes place is variable. Typic Cryandepts also occur on terraces bordering streams draining volcanic areas, where the coarser ash, because of its lightness, has been carried well beyond the transition zone in uplands. This alluvial ash is generally more tightly packed and has lower porosity than other ash deposits (Sokolov and Belousova, 1966). The principal soils associated with Typic Cryandepts are poorly drained Andeptic Cryaquepts and Histosols of depressions.

Dystric Cryandepts differ from soils of the Typic subgroup in that they consist dominantly of thixotropic volcanic material with high 15-bar water retention.

Base saturation, as determined with neutral normal ammonium acetate, in any layer between depths of 25 and 75 cm is not permitted to exceed 50%. A new subgroup of Eutric Cryandepts (or its equivalent in a revised taxonomy) may be needed to accommodate any cold Andepts that fail to meet this requirement. Although high percentages are not part of the definition of the subgroup, the organic matter content of most Dystric Cryandepts is very high and may approach or even exceed in some horizons the minimum percentages required for Histosols.

In most cases the soils are formed in fine-grained volcanic ash that has been deposited a substantial distance from the source and that has been subject to weathering for a considerable time. Also included in the subgroup are stratified coarse and fine ashy soils in which the fine ash with thixotropic properties occupies the greater part of the 25–100 cm section or in which the average 15-bar water retention for the section exceeds 20%. Some soils formed in strongly weathered basaltic rock with a high glass component are Dystric Cryandepts, but, more commonly, soils in basaltic detritus in cold areas are Mollisols. As in the Typic Cryandepts, a layer of unweathered recent ash less than 50 cm thick may overlie the soils.

Associated soils are Andeptic Cryaquepts, Histosols, and soils of other orders in uplands, including those in which volcanic deposits are less than 35 cm thick. Dystric Cryandepts are more susceptible to degradation to Spodosols in cool, humid climates with fairly warm summers than are Typic Cryandepts.

Lithic Cryandepts and *Dystric Lithic Cryandepts* overlie bedrock or other coherent material at depths of less than 50 cm. Properties of the soils are otherwise like those of the Typic and Dystric subgroups, respectively. These soils commonly are associated with rock outcrops or, especially at high elevations, with Cryofolists that consist of a thin layer of organic material over the bedrock.

Entic Cryandepts either have a surface deposit of unweathered ash more than 50 cm thick over an umbric or mollic epipedon, or have a cambic horizon but not enough accumulation of organic matter to satisfy the requirements for an umbric or mollic epipedon. In the latter case, they represent a transition stage between Entisols in fresh ash and either Typic or Dystric Cryandepts. Because the epipedons form quite rapidly in ashy materials, the soils are short-lived and, therefore, not extensive.

Some soils formed in material weathered from basic and ultrabasic rocks in the island of Rhum in the Inner Hebrides have the morphology of Andepts and include a thin discontinuous "ironpan" (Ragg and Ball, 1964), possibly a placic horizon. These soils also have high base saturation. It is possible that an additional subgroup defined as an intergrade to both the Eutrandepts and the Placandepts of temperate climates will be required to cover this situation. Ashy soils in very humid climates where placic horizons are likely to occur, however, are generally Spodosols in cold regions.

References

Besoain, E. (1968). Imogolite in volcanic soils of Chile. *Geoderma* 2, 151–169.

Broadbent, F. E., Jackman, R. H., and McNicoll, J. (1964). Mineralization of carbon and nitrogen in some New Zealand allophanic soils. *Soil Sci.* 98, 118–128.

Chichester, F. W., Youngberg, C. T., and Harward, M. E. (1969). Clay mineralogy of soils formed in Mazama pumice. *Soil Sci. Soc. Am. Proc.* 33, 115–120.

Espinoza, W., Rust, R. H., and Adams, R. S., Jr. (1975). Characterization of mineral forms in Andepts from Chile. *Soil Sci. Soc. Am. Proc.* 39, 556–561.

Everett, K. R. (1977). Soils. *In* "The Environment of Amchitka Island, Alaska." (M. L. Merritt and R. G. Fuller, eds.), pp. 179–202. U. S. Energy Res. and Dev. Adm., Washington, D.C.

Frei, E. (1978). Andepts in some high mountains of East Africa. *Geoderma* 21, 119–131.

Helgason, B. (1963). Basaltic soils of south-west Iceland, I. *J. Soil Sci.* 14, 64–72.

Helgason, B. (1968). Basaltic soils of south-west Iceland, II. *J. Soil Sci.* 19, 127–134.

Higashi, T., and Wada, K. (1977). Size fractionation, dissolution analysis, and infra-red spectroscopy of humus complexes in Ando soils. *J. Soil Sci.* 28, 653–663.

Karpachevskiy, L. O. (1965). Some features of soil formation in Kamchatka. *Sov. Soil Sci.,* 1259–1267.

Liverovskiy, Yu. A. (1971). Volcanic ash soils of Kamchatka. *Sov. Soil Sci.* 3, 271–278.

McIntosh, P. D. (1980). Weathering products in Vitrandept profiles under pine and manuka, New Zealand. *Geoderma* 24, 225–239.

Parfitt, R. L., Fraser, A. R., and Farmer, V. C. (1977). Adsorption on hydrous oxides. III. Fulvic acid and humic acid on goethite, gibbsite, and imogolite. *J. Soil Sci.* 28, 289–296.

Ragg, J. M., and Ball, D. F. (1964). Soils of the ultra-basic rocks of the Island of Rhum. *J. Soil Sci.* 15, 124–133.

Rieger, S. (1974). Humods in relation to volcanic ash in southern Alaska. *Soil Sci. Soc. Am. Proc.* 38, 347–351.

Rieger, S., and Wunderlich, R. E. (1960). "Soil Survey and Vegetation of Northeastern Kodiak Island, Alaska." U. S. Dept. Agr., Soil Surv. Ser. 1956, No. 17. Washington, D. C.

Rieger, S., Schoephorster, D. B., and Furbush, C. E. (1979). "Exploratory Soil Survey of Alaska." U. S. Dept. Agr., Soil Cons. Serv. Washington, D. C.

Simonson, R. W., and Rieger, S. (1967). Soils of the Andept suborder in Alaska. *Soil Sci. Soc. Am. Proc.* 31, 692–699.

Singer, M., and Ugolini, F. C. (1974). Genetic history of the well-drained subalpine soils formed on complex parent materials. *Can. J. Soil Sci.* 54, 475–489.

Sokolov, I. A., and Belousova, N. I. (1964). Organic matter in Kamchatka soils and certain problems of illuvial-humic soil formation. *Sov. Soil Sci.,* 1026–1035.

Sokolov, I. A., and Belousova, N. I. (1966). Water-physical properties and water-thermal regime of ochrous volcanic forest soils of Kamchatka. *Sov. Soil Sci.,* 533–543.

Sokolov, I. A., and Karayeva, Z. S. (1965). Migration of humus and some elements in the profile of volcanic forest soils of Kamchatka. *Sov. Soil Sci.,* 467–475.

Ulrich, H. P. (1946). Morphology and genesis of the soils of Adak Island, Aleutian Islands. *Soil Sci. Soc. Am. Proc.* **11**, 438–441.

Wada, K., and Harward, M. E. (1974). Amorphous clay constituents of soils. *Adv. Agron.* **26**, 211–260.

Wada, K., and Higashi, T. (1976). The categories of aluminum- and iron- humus complexes in Ando soils determined by selective dissolution. *J. Soil Sci.* **27**, 357–368.

Wada, K., and Kubo, H. (1975). Precipitation of amorphous aluminosilicates from solutions containing monomeric silica and aluminum ions. *J. Soil Sci.* **26**, 100–111.

Wells, N., and Furkert, R. J. (1972). Bonding of water to allophane. *Soil Sci.* **113**, 110–115.

Chapter 9

INCEPTISOLS: CRYOCHREPTS

Cryochrepts are well- or moderately well-drained Inceptisols with an ochric epipedon and a brown cambic horizon. Most uncultivated Cryochrepts have a thin 0 horizon made up of partially decomposed forest or tundra litter at the surface. Many have an A horizon that is not thick or dark enough to be a mollic or umbric epipedon, but in some the cambic horizon lies immediately below the 0 horizon. A thin lighter-colored horizon may occur between the 0 or A horizon and the cambic horizon. In soils that remain frozen well into the summer or that are underlain by permafrost, there may be mottles or organic streaks within the cambic horizon or in the soil below it. A horizon of calcium carbonate accumulation, in many cases substantial enough to constitute a calcic horizon, may occur below the cambic horizon.

I. Genesis of Cryochrepts

A. CONDITIONS OF FORMATION

Cryochrepts support either forest or tundra vegetation. In forested areas they form mostly in loamy materials under cold continental climates with mean annual precipitation less than 375 mm, most of which falls as showers in the summer months, and with mean summer air temperatures of 12°–16° C.

These are the dominant soils of well-drained uplands in interior Alaska (Rieger *et al.*, 1979), much of northwestern Canada (Clayton *et al.*, 1977), and the dry interior of eastern Siberia (Sokolov *et al.*, 1976). Under similar continental temperature regimes with higher precipitation rates, well-drained non-calcareous loamy soils are generally Spodosols. The forest in all cases consists of either coniferous or mixed coniferous and hardwood trees.

Cryochrepts may also form in young calcareous materials in areas with greater precipitation and less strongly continental climates. In those areas they are relatively short-lived and are soon converted to Alfisols.

In alpine and, to a lesser extent, arctic tundra areas, Cryochrepts occur where summers are fairly warm. In most tundra areas, however, Cryumbrepts or Cryorthods are the dominant well-drained soils. All of these soils are commonly underlain at some depth by permafrost and are associated with Cryaquepts and Histosols with shallow permafrost tables.

The parent materials of most Cryochrepts are of Holocene age. They include loess, material of alluvial origin, glacial till, and weathering products of rocks that were ice-covered during the Pleistocene era. They almost invariably have loamy textures. Cryochrepts cannot form in sandy materials because no cambic horizon can be recognized in sands, but they do exist where the soil has developed in a thin layer of loess overlying the sand. Well-drained, sandy soils with no loamy surface layer in areas dominated by Cryochrepts are either Entisols or Spodosols. Few clayey Cryochrepts exist. In forested areas Alfisols form fairly quickly in clayey materials, and in tundra areas most clayey soils have shallow permafrost tables and are poorly drained.

B. THE A HORIZON

The principal processes involved in the formation of forested Cryochrepts are the accumulation of organic litter at the surface and the release of iron and aluminum from primary minerals by weathering. Relatively small amounts of organic matter are added to these soils in the cold, dry environment in which they form, and decomposition rates are lower than in the warm fringes of the boreal belt. Wildfires that destroy part of the litter are common occurrences and further reduce the amount of organic matter that accumulates. 0 horizons are seldom more than 10 cm thick, and organic materials even in the lower part of the horizon are only partially decomposed. The root system in most of these soils is very shallow; only a few fine roots penetrate as deeply as 30 cm into the mineral soil. Under strongly continental climates, however, some fine roots may penetrate to considerable depth in cracks formed during extremely cold winters (Sokolov and Tursina, 1979). Earthworms and other soil fauna are few in number. Under these conditions, even in calcareous materials, the process of accumulation, decomposition, and dispersal of organic matter that results in formation of a mollic epipedon in somewhat warmer soils can operate only weakly. The A horizon that forms in Cryochrepts is not thick enough to be a mollic or umbric epipedon, and organic matter percentages in the horizon are lower than in most Cryoborolls or Cryumbrepts.

Because of the shallow root system, the consequent low rate of recycling of cations through the vegetation, and the generally acidic character of the litter, the 0 and A horizons of Cryochrepts are commonly moderately to strongly acid and are the most acidic horizons in the soil profile (Kuz'min and Chernegova, 1976). Weathering and release of iron and aluminum from primary minerals is most intense in the A horizon and small additional quantities may be brought to the surface in solution during summer dry periods and periods of very low winter temperatures (Ufimtseva, 1963). In the early development of the soils, little migrates downward. As a result the highest concentration of iron and aluminum in Cryochrepts, in contrast with Alfisols and Spodosols, is in the surface horizon (Blume and Schwertmann, 1969).

C. THE CAMBIC HORIZON

The cambic horizon seldom extends to depths greater than 60 cm. Its brown color is the result of the process of *brunification* (Duchafour and Souchier, 1978)—that is, the release of iron oxide by weathering of primary minerals or the decay of organic materials and its accumulation at the site of release. A high proportion of the total iron in the horizon is in the form of crystallized (dithionite-extractable) iron oxides that have developed in place and not from material translocated from the A horizon (Blume and Schwertmann, 1969). Similarly, most clay in the cambic horizon is not of illuvial origin but is a product of weathering. It occurs as minute floccules, commonly with thin coatings of iron oxide (Duchafour and Souchier, 1978), rather than as linings on peds or in pores (Sokolov *et al.,* 1976). Clay migration, although it does eventually take place, is inhibited by stable aggregation with organic matter in the surface horizon and by the low leaching potential of dry continental climates (Morozova, 1965). Leaching of soluble materials is common, however; horizons of calcium carbonate accumulation occur under cambic horizons of Cryochrepts, especially in soils developing in originally calcareous parent materials, and cambic horizons in noncalcareous materials may become strongly acid despite the low precipitation rate.

D. TRANSLOCATION OF SOLUBLE SUBSTANCES

In the cold, dry climate under which they form, the Cryochrepts can persist without significant change for long periods. If there are no additions of fresh loess or other sediments, however, moisture from summer rainfall percolating through the acid 0 horizon eventually modifies the soils. Soluble aluminum hydroxides migrate to the cambic horizon and manganese compounds to lower horizons. Less soluble iron compounds, some of them in combination with organic acids, migrate later and accumulate in small amounts in the cambic horizon or, in shallow soils, above the underlying bedrock (Mikhaylova, 1976). In calcareous soils, organic acids may move completely through the cambic horizon and accumulate in an underlying calcic horizon (Lutwick and Dormaar, 1973).

Soluble organic matter in forested Cryochrepts is dominantly humic acid in the upper horizons, but the small amount of organic matter that is translocated to greater depths is mostly fulvic acid (Fedorova and Yarilova, 1972; Kuz'min and Chernegova, 1976; Sokolov et al., 1976).

The translocation of small quantities of iron and organic matter from the surface horizon to the cambic horizon or deeper in the profile results in the development of a horizon of lighter color between the A horizon and the cambic horizon. Colors in this horizon may or may not have the low intensity (chroma) required for recognition of an albic horizon. The eluviated horizon can form even in parent materials that were originally calcareous, but it is most prominent in soils with acidic parent materials. The underlying cambic horizon in those soils gradually accumulates weakly crystalline (oxalate-extractable) and organically complexed (pyrophosphate-extractable) iron (Lutwick and Dormaar, 1973). In soils formed under spruce forest with a strongly acid 0 horizon, the lighter color may be in part the result of leaching of soluble silicon hydroxide from the 0 horizon and precipitation as silica in the upper mineral soil (Tyshkevich, 1958). Grayish spots in the lower part of the 0 horizon may also be a consequence of this process or may represent the onset of the spodic process (Ufimtseva, 1963).

E. TRANSLOCATION OF CLAY

Because only small quantities of material are translocated, there is little change in the cambic horizon during the initial formation of the light-colored horizon above it. As acidity increases, however, translocation of either clay or organometallic complexes occurs and the soils are gradually transformed to Alfisols or Spodosols. Where the parent materials are moderately saline, clay in the upper horizons may be dispersed under the influence of the sodium ion and may be translocated under alkaline rather than acidic conditions (Sokolov et al., 1976). Probably because of the inefficient leaching regime in most forested Cryochrepts, only those forming in the most acidic parent materials are eventually subjected to the spodic process. Most appear to be increasingly influenced by the process of argilluviation as they age and to be developing in the direction of Alfisols.

In coarse-silty soils an early indication of clay illuviation is the development of thin undulating bands within the cambic horizon that are higher in clay than the surrounding soil matrix (DeMent, 1962). Barring a change in climate, the Cryochrepts may never develop beyond this point. In coarse-loamy materials discontinuous coatings of oriented clay may form on sand grains (Morozova, 1965), and in finer materials clay coatings may occur sporadically on ped surfaces or in pores (Mikhaylova, 1976). In the latter case it may be difficult to determine whether the clay skins are the result of argilluviation or of freezing processes (Chapter 2). The occurrence of Alfisols on older surfaces in close geographical association with Cryochrepts on younger materials (Wright *et al.*, 1959), however, indicates that the final stage in the development of moderately acidic forested Cryochrepts is the formation of an argillic horizon.

Clays in all Cryochrepts are much like those of the parent materials. They are principally illites and smectites, with lesser amounts of chlorite and kaolinite (Day and Rice, 1964; Sokolov *et al.*, 1976).

F. DEVELOPMENT OF CRYOCHREPTS IN TUNDRA AREAS

In arctic and alpine tundra and tundra–forest transition areas, Cryochrepts occupy only positions with good surface drainage, such as ridge tops, escarpment edges, and steep southerly slopes. Most are gravelly (Ufimtseva, 1963; Day and Rice, 1964; Rieger *et al.*, 1979) and have ice-rich permafrost at depths greater than 1 m, but in some the permafrost table is shallower. In very dry areas there is no deformation of soil horizons because of heaving (Ufimtseva, 1963), but in more humid areas many of the soils are affected by cryoturbation, which results in barren spots on the surface, stone stripes, and other frost-action phenomena (Walmsley and Lavkulich, 1975). In some soils, especially those with fairly shallow permafrost tables, organic matter from upper horizons may accumulate immediately above the permafrost table as a result of frost-related soil movements (Pawluk and Brewer, 1975; Schnitzer and Vendette, 1975). In tundra areas either fulvic acid (Mikhaylov, 1960) or humic acid (Schnitzer and Vendette, 1975) may dominate in the soluble organic fraction, but the humic acid is poorly developed and has low resistance to degradation.

Cryochrepts of tundra areas form in many kinds of parent material, ranging from acidic to calcareous. Most of them are associated with more fully developed soils—Cryorthods, Cryumbrepts, or Cryoborolls—and probably represent an intermediate stage in development toward one of those groups. Because of both stable aggregation in horizons containing organic matter and disruptions caused by frost action, there is seldom any indication of clay illuviation or the development of an argillic horizon in these soils (Day and Rice, 1964; Pawluk and Brewer, 1975; Walmsley and Lavkulich, 1975). In some cases tundra Cryochrepts are relicts of formerly existing Cryumbrepts or

Cryoborolls from which the A horizon has been removed during a period of intense frost action. The solum in most of these soils is less than 30 cm thick.

II. Classification of Cryochrepts

Typic Cryochrepts are those with mean annual soil temperatures higher than 0° C, with no mottling except at depths well below the solum, and with at least one subhorizon in the upper 75 cm of the profile in which base saturation (at pH 7) is 60% or higher. They may have a light-colored horizon above the cambic horizon but have no indications of clay illuviation in the form of clayey lamellae within the cambic horizon. Depth to bedrock or other impervious substratum is greater than 50 cm, and no layer of volcanic ash in the upper part of the soil is as thick as 18 cm.

Most Typic Cryochrepts are forested, but some on steep, south-facing slopes (in the northern hemisphere) in the driest continental areas have a sparse brush (including sagebrush) and grass cover. Many have high base saturation and pH values close to neutrality throughout the profile. In others calcium, magnesium, and other exchangeable cations have been leached from the solum to the extent that base saturation is fairly low and the soil reaction is acidic, but because leaching effectiveness is low, they have accumulated in a horizon below the solum, but within 75 cm of the surface. In shallow loessial soils this accumulation commonly occurs at the base of the loess where it overlies gravelly or stony material.

Common associated soils that occupy lower slopes, depressions, and, in hilly areas, polar-facing slopes are Aquic Cryochrepts, Cryaquepts, and Histosols. At higher elevations where summer soil temperatures are lower and soils are generally more stony, the Cryochrepts grade to Cryorthods, Cryorthents, and Cryaquepts. The associated Cryaquepts and Histosols generally have thick 0 horizons and shallow permafrost tables.

Dystric Cryochrepts differ from the Typic subgroup in that base saturation throughout at least the upper 75 cm of the profile is less than 60%. These soils form primarily in acidic parent materials that are high in iron and aluminum (Douchafour and Souchier, 1978) and in areas where annual precipitation and leaching effectiveness approach levels that favor the development of Spodosols. The vegetation is generally coniferous forest (Tyshkevich, 1958). Organic matter levels in the Dystric Cryochrepts are higher than in the Typic subgroup, and a higher proportion of the iron and aluminum in the cambic horizon is complexed with organic matter and extractable by pyrophosphate (Lutwick and Dormaar, 1973). Although summer dry periods interfere with the operation of the spodic process, it is likely that in time a spodic horizon will develop in these soils.

Alfic Cryochrepts have thin clayey lamellae in the cambic horizon, but the combined thickness of the lamellae is not large enough to satisfy the thickness requirement for an argillic horizon. The clayey bands occur most commonly in soils formed in coarse-silty loess. They are believed to have originated as zones of slightly higher clay concentration in the original deposit. Over fairly long periods of time, the zones are concentrated into thin bands by freezing pressures, and the bands are later thickened by additions of small amounts of clay and organic matter from percolating water (DeMent, 1962). Clay skins on the faces of the fine, blocky peds within the bands and thin, bleached subhorizons above the bands apparently caused by lateral movement of perched water support this hypothesis. Continuing additions of clay to the bands may eventually result in coalescence to form an argillic horizon, but because of the low clay content of the original loess, this would be a very slow process at best.

The clayey lamellae occur only in material that has been stable with no new additions for thousands of years. Apart from the lamellae these soils are identical with Typic Cryochrepts and support the same vegetation. Associated soils are like those that border soils of the Typic subgroup.

Andic Cryochrepts differ from soils of the Typic subgroup in having a layer of volcanic ash at least 18 cm thick at the surface. In a large area of east central Alaska and adjoining portions of the Yukon Territory covered by an ash deposit that is in places thicker than 18 cm, the ash is essentially unaltered about 1400 years after its deposit and the soil beneath the ash is the same as the Typic Cryochrepts (Rieger *et al.*, 1979). In the Rocky Mountains of Alberta, the ash in the upper part of the profile has weathered to a reddish color, and the soil beneath the ash has a higher proportion of amorphous (pyrophosphate-extractable) iron and aluminum than is normal for Cryochrepts (Pawluk and Brewer, 1975).

Lithic Cryochrepts have bedrock or other consolidated underlying material within 50 cm of the surface of the mineral soil. Mean annual soil temperatures may be higher or lower than 0° C, and volcanic ash deposits may or may not occur above the solum. These soils exist mostly on high ridges in hills and mountains and may support either forest or tundra vegetation. Where they occur on steep slopes, substantial amounts of organic matter may be incorporated in the soil as a result of downslope creep (Walmsley and Lavkulich, 1975). Illuviated iron may accumulate above the bedrock in stable positions (Mikhaylova, 1976).

Aquic Cryochrepts are moderately well-drained soils that are mottled in or below the cambic horizon or, in some instances, close to the surface (Fedorova and Yarilova, 1972). Fine iron–manganese–humus concretions may occur in the upper horizons as a result of intermittent periods of saturation (Mikhaylova, 1976). The vegetation is commonly a forest of slow-growing conifers. Aquic Cryochrepts occupy shallow depressions, sites bordering organic soils, gentle slopes at the foot of hills, or high areas with cooler summers than in areas dominated by Typic or Alfic Cryochrepts. In all of these positions, the soils receive or retain more moisture than associated well-drained soils and remain frozen well into the summer. Although they lose their excess moisture after complete thawing in the latter part of the summer, they have fairly long periods of saturation. Many are subject to intense cryoturbation and have patches and streaks of organic matter throughout the upper part of the soil.

These soils are similar to Aeric Cryaquepts and are distinguished from them chiefly by a more intense color (higher chroma) in the cambic horizon. They also resemble Aquic Cryorthents, which have only slight mottling and chromas that are not quite high enough to permit recognition of a cambic horizon. Where there have been no wildfires for a long time, the forest may become very dense, and the 0 horizon may grow thick enough to prevent complete thawing of the soil. Under these conditions the soil above the permafrost layer thus developed is converted to a Pergelic Cryochrept, or eventually, to a strongly mottled Pergelic Cryaquept. Removal of the forest by burning or clearing can result in reversion to the original condition. It is not uncommon for areas of both of these soils to exist in close association with Aquic Cryochrepts.

Pergelic Cryochrepts have mean annual soil temperatures of 0° C or less, but in all cases the permafrost table is well below the lower limit of the solum. Many that have formed in gravelly materials, especially in alpine tundra areas, have a zone of dry permafrost (temperatures perennially at or below 0° C, but no ice in the interstices between particles) that may overlie a deeper zone of ice-rich permafrost. Where ice-rich permafrost is at moderate depths, there may be mottling below the solum and a slight accumulation of organic matter at the permafrost table (Morozova, 1965; Pawluk and Brewer, 1975; Schnitzer and Vendette, 1975), but many Pergelic Cryochrepts in areas with low rainfall do not exhibit these characteristics (Sokolov *et al.,* 1976).

Pergelic Cryochrepts form under both boreal forest and shrub tundra vegetation. In forested soils of dry continental areas, surface moisture does not penetrate to the permafrost table, and the soils develop just as if there were no perennially frozen soil beneath them. Soil-forming processes and solum development are identical with those in nonpergelic, forested Cryochrepts, and there is no deformation of genetic horizons (Ufimtseva, 1963; Morozova, 1965). Many of these soils—in the Central Yakutian Lowland of Siberia, for example—have been farmed successfully and may even suffer from occasional droughtiness. A hazard in the cultivation of these soils is thermokarst pitting that may accompany recession of the permafrost table after clearing.

In tundra areas and in the transition zone between tundra and forest, where summers are cooler and the soils are generally moister, frost action may create a patterned or mounded surface. In the tundra, especially, there may be barren spots where the vegetation has been disrupted. Although disturbances are intermittent, separated by stable periods during which development is uninterrupted (Pawluk and Brewer, 1975), the cambic horizon may be partially destroyed. In strongly mounded areas, remnants of the cambic horizon may be interspersed within the mounds with C horizon material (Pettapiece, 1975). A new subgroup of *Pergelic Ruptic-Entic Cryochrepts* may be required for these disturbed soils. Pergelic Cryochrepts of tundra areas commonly occur as part of an association that also includes Pergelic Cryorthents, Cryorthods, Cryumbrepts, or Cryoborolls.

References

Blume, H. P., and Schwertmann, U. (1969). Genetic evaluation of profile distribution of aluminum, iron, and manganese oxides. *Soil Sci. Soc. Am. Proc.* **33**, 438–444.

Clayton, J. S., Ehrlich, W. A., Cann, D. B., Day, J. H., and Marshall, I. B. (1977). ''Soils of Canada.'' 2 vol., maps. Res. Branch, Can. Dept. Agr., Ottawa.

Day, J. H., and Rice, H. M. (1964). The characteristics of some permafrost soils in the Mackenzie Valley, N. W. T. *Arctic* **17**, 223–236.

DeMent, J. A. (1962). ''The Morphology and Genesis of the Subarctic Brown Forest Soils of Central Alaska.'' Unpub. Ph. D. Dissertation, Cornell U.

Duchafour, P., and Souchier, B. (1978). Role of iron and clay in genesis of acid soils under a humid, temperate climate. *Geoderma* **20**, 15–26.

Fedorova, N. M., and Yarilova, E. A. (1972). Morphology and genesis of prolonged seasonally frozen soils of western Siberia. *Geoderma* **7**, 1–13.

Kuz'min, V. A., and Chernegova, L. G. (1976). Brown Mountain-Taiga and Sod Forest soils in the Baykal region. *Sov. Soil Sci.* **8**, 373–380.

Lutwick, L. E., and Dormaar, J. F. (1973). Fe status of Brunisolic and related soil profiles. *Can. J. Soil Sci.* **53**, 185–197.

Mikhaylov, I. S. (1960). Some features of Sod Arctic soils on Bolshevik Island. *Sov. Soil Sci.*, 649–657.

Mikhaylova, R. P. (1976). Micromorphological and chemical characteristics of coarse humic Brown soils in the central mountain belt of the central Ural. *Sov. Soil Sci.* **8**, 249–256.

Morozova, T. D. (1965). Micromorphological characteristics of Pale Yellow Permafrost soils of central Yakutia in relation to cryogenesis. *Sov. Soil Sci.*, 1333–1342.

Pawluk, S., and Brewer, R. (1975). Investigation of some soils developed in hummocks of the Canadian sub-arctic and southern-arctic regions: 2. Analytical characteristics, genesis, and classification. *Can. J. Soil Sci.* **55**, 321–330.

Pettapiece, W. W. (1975). Soils of the subarctic in the lower Mackenzie basin. *Arctic* **28**, 35–53.

Rieger, S., Schoephorster, D. B., and Furbush, C. E. (1979). ''Exploratory Soil Survey of Alaska.'' U. S. Dept. Agr., Soil Cons. Serv. Washington, D. C.

Schnitzer, M., and Vendette, E. (1975). Chemistry of humic substances extracted from an arctic soil. *Can. J. Soil Sci.* **55**, 93–103.

Sokolov, I. A., and Tursina, T. V. (1979). Pale Yellow-Gray soils of central Yakutia, an analog of Gray Forest soils. *Sov. Soil Sci.* **11**, 125–136.

Sokolov, I. A., Naumov, Ye. M., Gradusov, B. P., Tursina, T. V., and Tsyurupa, I. G. (1976). Ultracontinental taiga soil formation on calcareous loams in central Yakutia. *Sov. Soil Sci.* **8**, 144–160.

Tyshkevich, G. L. (1958). Soils under spruce forests of the Carpathians. *Sov. Soil Sci.*, 136–138.

Ufimtseva, K. A. (1963). Mountain taiga soils of the Transbaikal region. *Sov. Soil Sci.*, 241–248.

Walmsley, M. E., and Lavkulich, L. M. (1975). Landform–soil–vegetation–water chemistry relationships, Wrigley area, N. W. T.: II. Chemical, physical, and mineralogical determinations and relationships. *Soil Sci. Soc. Am. Proc.* **39,** 89–93.

Wright, J. R., Leahey, A., and Rice, H. M. (1959). Chemical, morphological and mineralogical characteristics of a chronosequence of soils on alluvial deposits in the Northwest Territories. *Can. J. Soil Sci.* **39,** 32–43.

Chapter 10

INCEPTISOLS: CRYUMBREPTS

Cryumbrepts are well- or moderately well-drained Inceptisols with an umbric epipedon that have developed in materials other than volcanic ash. The umbric epipedon may overlie a brown cambic horizon similar to that in the Cryochrepts, or it may rest directly on largely unaltered parent material. Some Cryumbrepts are mottled at depths greater than 50 cm. Those that freeze deeply or are underlain by permafrost may have streaks of organic matter in the soil below the epipedon.

I. Genesis of Cryumbrepts

A. CONDITIONS OF FORMATION

Most Cryumbrepts form in acidic parent materials. Only rarely do they occur in originally basic materials that have been strongly leached (Brown, 1966). They support either forest, grassy, or tundra vegetation. Most forested Cryumbrepts are in areas with a cold maritime or subalpine climate in which precipitation rates are higher than in strongly continental areas but in which summer temperatures are low enough to inhibit the spodic process (Assing, 1960; Nimlos and McConnell, 1965). They also occur on colluvial slopes in continental areas where there is a continuing deposition of organic-rich material. The forest may consist entirely of conifers or may be mixed conifers and birch. In many cases the forest is open and grasses dominate the ground cover. In subalpine areas, particularly, grassy meadows rather than trees may make up the vegetation.

Crumbrepts are the dominant well-drained soils in the arctic tundra and in higher and colder alpine tundra areas. In the arctic they and other well-drained soils occupy only a very small proportion of the total area, but in some areas of high alpine tundra, nearly one-half of the soils may be Cryumbrepts (Retzer, 1965). Soils formed in calcareous and other basic materials in these areas are dominantly Cryoborolls, however, and in somewhat warmer tundra areas, the Cryumbrepts give way to Cryorthods. Cryorthods rather than Cryumbrepts may also develop in colder areas where parent materials are coarse and acidic, as in granitic detritus (Sokolova, 1964; Brown, 1966). The vegetation on Cryumbrepts in the tundra consists primarily of low shrubs, grasses, and lichens, in sharp contrast to the sedges and mosses that dominate the cover on associated poorly drained soils.

B. THE UMBRIC EPIPEDON

Slow accumulation of organic matter in the upper part of the soil is the principal process in the development of Cryumbrepts. O horizons are commonly thicker than in the Cryochrepts, and because of the greater available moisture, root systems are denser. As in the Cryoborolls, additions of organic matter to the upper part of the mineral soil come about through decomposition of roots, faunal activity, and, to some extent, frost action. Decomposition rates are lower in the Cryumbrepts, however, because of the lower nutrient status of the parent materials and the consequent lower biological activity. Resistance to decomposition in these soils is indicated by a low cation exchange capacity of the organic matter (Karavayeva, 1958) and by narrow carbon to nitrogen ratios (Karavayeva and Targul'yan, 1963; Nimlos and McConnell, 1965). The umbric epipedon, as a result, forms more slowly and is generally thinner than the mollic epipedon of the Cryoborolls. Organic matter percentages decrease rapidly and steadily with depth (Retzer, 1956; Drew and Tedrow, 1957; Rode and Sokolov, 1960; Karavayeva and Targul'yan, 1963; Brown, 1966), and there is little translocation of organic compounds in the profile (Brown and Tedrow, 1964; Rieger, 1966) other than by cryoturbation. Organic stains have been observed on the undersides of stones in some coarse-textured Cryumbrepts (Retzer, 1956; Nimlos and McConnell, 1965), but it is likely that they owe their existence to the concentration of roots under the stones where the moisture supply is greatest (Retzer, 1965).

In many Cryumbrepts fulvic acid is the dominant extractable fraction of organic matter throughout the profile (Assing, 1960; Rode and Sokolov, 1960).

Others, especially those with a high proportion of lichens in the vegetation, are high in humic acid in the umbric epipedon although fulvic acid is dominant at depth (Karavayeva, 1958).

Most Cryumbrepts have a brown cambic horizon immediately below the umbric epipedon, but many, especially those formed in sandy materials, have no cambic horizon. As in the Cryochrepts, the brown color of the cambic horizon is principally attributable to iron oxides released by weathering from primary minerals. Some translocation of iron, along with aluminum and manganese, apparently does take place in many Cryumbrepts. Although in the youngest soils, the concentration of extractable iron is highest in the upper horizons because of more intense weathering and biological activity at the surface (Rode and Sokolov, 1960; Nimlos and McConnell, 1965), more strongly developed Cryumbrepts may have higher concentrations in the lower part of the umbric epipedon (Rieger, 1966) or in the underlying cambic horizon (Drew and Tedrow, 1957; Rode and Sokolov, 1960). Iron incrustations on the undersides of stones have been observed in Cryumbrepts of high mountains (Karavayeva, 1958; Sokolova, 1964). Migration in the form of organometallic complexes, as in the Spodosols, apparently is not important in these soils, but there are indications that some Cryumbrepts in depressed or protected sites where snow accumulates, soil temperatures are somewhat higher, and leaching effectiveness is greater, will eventually be subject to the spodic process. Some such soils in the tundra of northern Alaska, for example, exhibit weakly developed bleached horizons below the umbric epipedon in which chlorite has been replaced by montmorillonite in the clay fraction—a characteristic feature of the albic horizon of Spodosols—although spodic horizons have not yet formed (Brown and Tedrow, 1964).

No translocation of clay in appreciable amounts is likely to take place in these strongly acidic, continually moist soils. Most Cryumbrepts have coarse-loamy or coarser textures, and many, especially in tundra areas, are gravelly or stony. The proportion of silt and clay is generally highest in the surface horizons where weathering is most intense (Retzer, 1956) and where, in many cases, there are additions of windlaid material (Ugolini, 1966a).

II. Classification of Cryumbrepts

Typic Cryumbrepts are defined as well-drained soils with no mottling except at depths below 75 cm, with both an umbric epipedon and an underlying cambic horizon and with mean annual soil temperature higher than 0° C. Depth to bedrock or other impervious substratum is greater than 50 cm, and no layer of volcanic ash as thick as 18 cm occurs in the upper part of the soil.

Except in high mountains at low latitudes, Typic Cryumbrepts are either forested or grass-covered. Most are strongly acidic and have low base saturation to great depth, but base saturation increases with depth. At low latitudes they grade to Boralfs or Borolls (Assing, 1960) or, in some cases, to Orthods (Butuzova, 1966) at lower elevations. At higher latitudes they are associated with Pergelic Cryaquepts and Histosols in poorly drained positions and grade to Cryorthods and Cryochrepts in the direction of greater continentality of climate.

Entic Cryumbrepts differ from soils of the Typic subgroup only in that the cambic horizon is absent. In some of these soils on colluvial slopes, the umbric epipedon is created by additions of organic-rich materials from higher elevations, and there is little development in place. In others, the soil below the umbric epipedon is too sandy to permit recognition of a cambic horizon.

Andic Cryumbrepts have a layer of volcanic ash at least 18 cm thick in the upper part of the soil. Where there has been substantial weathering of the ash, the soils have many of the characteristics of Andepts, including an amorphous clay component and a high organic matter percentage in the umbric epipedon. At the other extreme there may be simply a surface deposit of relatively unaltered ash over a buried umbric epipedon. The soils may or may not have a cambic horizon.

Aquic Cryumbrepts are mottled between depths of 50 and 75 cm as a result of intermittent saturation, generally over a layer that remains frozen well into the summer months. These soils occur most commonly under alpine meadow vegetation.

Pergelic Cryumbrepts have mean annual soil temperatures of 0° C or lower and, therefore, permafrost at some depth. Virtually all Cryumbrepts under arctic tundra and many under high alpine tundra vegetation are in this subgroup, as well as some forested Cryumbrepts at the cold fringe of the boreal belt or on steep, polar-facing slopes at somewhat lower latitudes. In tundra areas the soils form only in relatively coarse-textured materials in exceptionally well-drained positions such as ridge tops or narrow belts immediately above steep escarpments. The solum is seldom thicker than 30 cm and becomes progressively thinner with increasing altitude or latitude (Tedrow and Brown, 1962; Brown, 1966). The A horizon is normally thin, and an umbric epipedon can be recognized only by averaging color values over the upper 18 cm of the profile. A brown cambic horizon commonly, but not always, underlies the A horizon.

The permafrost table in most of these soils is well over 1 m deep and has little direct effect on soil development, but many of the soils are nevertheless subject to frost movements. In some soils surficial cracking can increase root penetration to the extent that the A horizon thickens and obliterates any cambic horizon that may have formed (Ugolini, 1966a). In others disturbances can destroy the umbric epipedon completely and convert the soil to a Cryorthent or Cryochrept or destroy it only partially to create complex pedons in which the umbric epipedon remains on polygon borders but is absent in polygon centers. Soils in which the umbric epipedon is discontinuous are properly classified as *Pergelic Ruptic-Entic Cryumbrepts*. * Because a long period of stability is necessary for development of the solum, ruptic Cryumbrepts can form only where severe frost movements are rare and occur no more than sporadically after unusual events such as a year with extremely high rainfall (Ugolini, 1966b) or an exceptionally large number of summer freeze-thaw cycles.

The dominant soils in most areas where Pergelic Cryumbrepts occur are Pergelic and Histic Pergelic Cryaquepts. Well-drained, associated soils are Pergelic Cryochrepts and Cryorthents and, in the warmest sections of the arctic tundra and in alpine areas, Pergelic Cryorthods.

Lithic Cryumbrepts overlie bedrock at depths of less than 50 cm. They may or may not have a cambic horizon and may have temperatures either above or below 0° C. The vegetation may be either forest, meadow, or tundra. Where bedrock is exposed at the surface in part of each pedon, the soils are classified as *Ruptic-Lithic Cryumbrepts*. Where frost heaving has disrupted the soil so that an umbric epipedon occurs only in part of the pedon, they are *Lithic Ruptic-Entic Cryumbrepts*. These soils occur principally in alpine areas.

References

Assing, I. A. (1960). Some features of humus formation in mountain soils of northern Tyan'-Shan'. *Sov. Soil Sci.,* 1277–1282.

Brown, J. (1966). "Soils of the Okpilak River Region, Alaska." U. S. Army Cold Regions Res. Eng. Lab. (CRREL) Res. Rep. 188. Hanover, New Hampshire.

Brown, J., and Tedrow, J. C. F. (1964). Soils of the northern Brooks Range, Alaska: 4. Well-drained soils of the glaciated valleys. *Soil Sci.* **97,** 187–195.

Butuzova, O. V. (1966). Genesis of humic Cryptopodzolic soils. *Sov. Soil Sci.,* 485–490.

Drew, J. V., and Tedrow, J. C. F. (1957). Pedology of an Arctic Brown profile near Point Barrow, Alaska. *Soil Sci. Soc. Am. Proc.* **21,** 336–339.

Karavayeva, N. A. (1958). High-mountain soils of the eastern Sayans. *Sov. Soil Sci.,* 397–401.

Karavayeva, N. A., and Targul'yan, V. O. (1963). Contribution to the study of soils on the tundras of northern Yakutia. *In* "Soils of Eastern Siberia" (Ye. N. Ivanova, ed.), pp. 57–78. U. S. Dept. Comm., Clearinghouse for Fed. Sci. and Tech. Inf., Trans. TT 69–55073, Springfield, Virginia.

Nimlos, T. J., and McConnell, R. C. (1965). Alpine soils in Montana. *Soil Sci.* **99,** 310–321.

Retzer, J. L. (1956). Alpine soils of the Rocky Mountains. *J. Soil Sci.* **7,** 22–32.

---

* This subgroup has not yet been recognized in the United States Taxonomy.

Retzer, J. L. (1965). Present soil-forming factors and processes in arctic and alpine regions. *Soil Sci.* **99**, 38–44.

Rieger, S. (1966). Dark well-drained soils of tundra regions in western Alaska. *J. Soil Sci.* **17**, 264–273.

Rode, T. A., and Sokolov, I. A. (1960). The Transbaykal mountain-tundra landscapes. *Sov. Soil Sci.*, 384–391.

Sokolova, T. A. (1964). Effect of rocks on Podzol formation. *Sov. Soil Sci.*, 233–240.

Tedrow, J. C. F., and Brown, J. (1962). Soils of the northern Brooks Range, Alaska: Weakening of the soil-forming potential at high arctic altitudes. *Soil Sci.* **93**, 254–261.

Ugolini, F. C. (1966). Soils of the Mesters Vig District, northeast Greenland: I. The Arctic Brown and related soils. *Medd. om Gron., Bd. 176, No. 1,* 22 pp. (a)

Ugolini, F. C. (1966). Soils of the Mesters Vig District, northeast Greenland: II. Exclusive of Arctic Brown and Podzol-like soils. *Medd. om Gron., Bd. 176, No. 2,* 25 pp. (b)

Chapter 11

INCEPTISOLS:
CRYAQUEPTS

Aquepts are Inceptisols that are saturated with oxygen-depleted water during much or all of the thaw period and that have characteristics induced by reducing conditions during the periods of saturation. Excluded from Inceptisols and therefore from Aquepts are poorly drained soils with a mollic epipedon (except for soils formed in volcanic ash and soils in which an underlying horizon is low in bases), sandy or stratified soils in which no cambic horizon can be recognized (except for poorly drained soils with a mean annual temperature of $0°$ C or less), and soils with a spodic or an argillic horizon. Cryaquepts commonly occur, however, in association with well-drained soils that have these diagnostic subsurface horizons and in the wettest parts of landscapes that are dominated by poorly drained soils with mollic epipedons (Everett and Parkinson, 1977).

Cryaquepts are widespread in cold regions and are the dominant soils in the tundra and the coldest parts of the boreal belt. In those areas virtually all of the Cryaquepts, along with most of the associated well-drained soils, have perennially frozen substrata. In continental areas with warmer summers, Cryaquepts with permafrost are commonly associated with well-drained non-pergelic soils. At the warm fringes of the boreal belt and in most forested areas with cool maritime climates, the Cryaquepts have no permafrost.

I. Genesis of Cryaquepts

A. CONDITIONS OF FORMATION

Poorly drained soils in cold regions, even more than in other regions, are not confined to depressions or other areas with shallow water tables. They also occupy many level upland areas, steep, polar-facing slopes, slopes subject to seepage, and, especially in arctic tundra areas, rolling or hilly uplands—that is, any position where moisture in excess of field capacity accumulates because of a high water table, slow subsoil permeability, or a frozen substratum during all or a substantial portion of the summer. Cryaquepts form wherever stagnant water fills interstices between soil particles most of the time that the soil is not frozen and where there is no opportunity for free drainage and the development of spodic or argillic horizons. The only diagnostic horizons that can form in such soils are umbric, mollic, or histic epipedons at the surface and dull-colored or mottled cambic horizons in the subsoil.

Water-tolerant plants dominate the vegetation in Cryaquepts, but the range in vegetation types is large. Common trees are black spruce, cedar, lodgepole pine, larch or tamarack, alder, and willows. A thick ground cover of shrubs, grasses, sedges, and mosses occupies the forest floor. In tundra areas sedges and mosses are dominant but are accompanied by lichens and a wide variety of shrubs including willows, alder, and dwarf birch. Tussock sedges are prominent in most tundra uplands but seldom occur in the lowest and wettest areas. In the colder parts of the boreal belt, many Cryaquepts support tundra vegetation, in contrast with the forest on adjacent well-drained soils.

B. THE GLEYING PROCESS

The process responsible for the dull gray and commonly mottled cambic horizon that is the principal characteristic of Cryaquepts is *gleization* or *gleying.* It takes place in soils with virtually stagnant moisture regimes (Archegova, 1974) and is equally effective under tundra and forest vegetation (Leahey, 1947; Rode and Sokolov, 1960). When moisture in larger quantities than can be adsorbed by soil particles fills interstices between the particles and prevents air from entering the soil for long periods, oxygen dissolved in the water is depleted. Under the reducing environment thus created, iron is converted to the ferrous form, and soil colors become predominantly dull gray or may take on a greenish or bluish cast. Some of the ferrous iron may be released from ferrous minerals (McKeague, 1965a), but most of it forms as a direct result of the activities of iron-reducing microorganisms (Daragan, 1967). This is demonstrated by the almost complete cessation of the reduction process after sterilization of the soil and intensification of the process on addition of glucose to oxygen-depleted soils (Nepomiluyev and Kozyrev, 1970). Only in soils, such as those of coastal meadows, in which parent materials were deposited under water and have never been aerated is the high ferrous iron concentration and the associated soil color an inherited soil characteristic.

The gleying process requires the presence of organic material and microflora capable of decomposing the organic matter under anaerobic conditions and of transforming iron from the ferric to the ferrous state. Organic acids released from decomposing organic matter or secreted by organisms bring iron from primary or secondary minerals into solution where it is reduced by bacterial action (Daragan, 1967). The ferrous iron introduced into the soil solution during this process is the immediate cause of the colors associated with gleying.

C. BIOLOGICAL ACTIVITY AT LOW TEMPERATURES

It is commonly held that biological activity ceases at temperatures below 5° C, and this concept enters into several definitions in the United States Taxonomy, including that of the aquic moisture regime. A number of studies in tundra soils, however, have shown that indigenous microflora, including bacteria, molds, actinomycetes, and algae are capable of active growth at temperatures as low as 2° C (Sushkina, 1960; Cameron *et al.,* 1970; Boyd and Boyd, 1971). Vigorous roots of tundra sedges have been observed immediately above the permafrost table at temperatures below 1° C (Billings *et al.,* 1977). Although the number of microorganisms and the rate of decomposition in tundra soils is lower than in warmer soils, they may still contain as many as several million organisms per gram of soil at low temperatures (Ivarson, 1965). The number decreases with depth in the very cold soils (Jordan *et al.,* 1978), but living bacteria are present even in perennially frozen soil (Ivarson, 1965; Cameron *et al.,* 1970). The organisms involved in decomposition of organic matter at low temperatures are primarily fungi and anaerobic bacteria.

D. ORGANIC MATTER

Because the poorly drained soils are wet and cold during much of the summer, organic matter accumulates at the surface in larger quantity than in most associated well-drained soils. Much of it remains undecomposed. In most soils decomposition products of the fraction that does decay have an acidic reaction, and the soluble portion is largely fulvic acid (Kreida, 1958; Ignatenko, 1967b). In calcareous soils, however, there is greater humification of the 0 horizon (Pettapiece, 1975) and a higher proportion of humic acid (Bunting and Hathout, 1971). Eventually this results in the development of a mollic epipedon and classification of these soils as Cryaquolls rather than Cryaquepts.

Although most of the organic matter is contained in the 0 horizon and the uppermost mineral horizons, acidic Cryaquepts are moderately high in organic matter to considerable depth. The organic matter at depth is largely derived from plant roots and material from the surface horizon that has been displaced by frost-related turbation. Much of the buried organic matter is not completely decomposed because of the generally low temperature and the relatively small number of active microorganisms at depth (Ignatenko, 1971), but in addition to this relatively inert material, small quantities of organic acids that are the soluble products of decomposition at the surface migrate to lower horizons and permeate much of the soil (Kreida, 1958; Karavayeva and Targul'yan, 1963; Archegova, 1974). The mobility of these acids is greatest in acidic soils; it is drastically curtailed by excess calcium in the upper horizons (Pettapiece, 1975). In some arctic areas where the formation and drainage of lakes is a recurring phenomenon, the parent material itself may be the organic-rich bed of a former lake (Tedrow, 1969).

In many soils with permafrost that have fairly high concentrations of organic matter throughout the profile, humus coloration is noticeable only in the upper horizons and in a layer at the permafrost table (Karavayeva and Targul'yan, 1960; Pettapiece, 1974). Much of the organic matter in the intermediate horizons and under bare spots is colorless (Ignatenko, 1967b; Karavayeva, 1973, 1974), mainly because of low mineralization of organic acids in the cold environment of the soils (Ignatenko, 1971). The organic matter content under barren spots in frost-stirred soils is commonly as high as under the vegetation (Karavayeva, 1963; Ignatenko, 1966).

E. IRON DISTRIBUTION

The movement of iron in Cryaquepts is complicated by reactions with organic acids and by alternations between the ferrous and ferric states. In upland gleyed soils of humid areas that are only periodically saturated to the surface, the greatest loss of iron is from the mineral soil just below the 0 horizon, where biological activity is greatest and the largest amounts of ferrous iron are released. The highly soluble ferrous iron moves downward or laterally to lower topographic positions and leaves an upper horizon that, because of its low iron content, has few or no mottles (McKeague, 1965b; Ignatenko, 1966). Some of the iron may be deposited in a band at the boundary between the zone with seasonal aerobic condition and the zone of permanent saturation (Everett and Parkinson, 1977). Bands of this kind differ from spodic or placic horizons in that the iron is not associated with organic matter or dithionite-extractable aluminum (McKeague, 1965a).

In most acidic Cryaquepts, however, the highest concentrations of extractable iron are in surface horizons that are high in organic matter (Kreida, 1958; Rubtsov, 1964; Ivanova, 1965; Bel'chikova, 1966). In addition to iron that is associated with organic matter, the concentration of iron in the upper horizons is increased by the upward migration and oxidation of ferrous iron during dry summer periods, especially in the most poorly drained soils (Holowaychuk *et al.*, 1966). A secondary iron maximum occurs at the permafrost table in soils with an accumulation of organic matter at that depth (Ignatenko, 1971). Iron is evenly distributed in other horizons (Karavayeva, 1963; Ignatenko, 1966; Karavayeva, 1974; Pettapiece, 1974). Fulvic acid and iron combine, as in the Spodosols, to form complex organometallic compounds (Kreida, 1958; Moore, 1973; Pettapiece, 1974), but there normally is no single horizon in which most of those compounds precipitate. This is in large part due to the absence of a consistent leaching regimen and to the alternation between reducing and oxidizing conditions over the course of the summer. Because iron is also mobile in the ferrous state, independently of organic acids, the correlation between the distribution of iron and organic matter in the soil is not as strong as in the Spodosols (Karavayeva and Targul'yan, 1960). Like iron, aluminum tends to concentrate in the surface horizon of saturated cold soils. In soils with an accumulation of organic matter at the permafrost table, however, there is no corresponding increase in the aluminum concentration (Ignatenko, 1971).

F. MOTTLING

The dull colors that typify gleyed soils are most pronounced in horizons that are permanently saturated or that remain saturated over most of the summer. Continual saturation is generally owing to a shallow groundwater table or to water perched above ice-rich permafrost (Karavayeva, 1963). Reducing conditions are maintained in these saturated zones by biological destruction of soluble organic acids that diffuse into them from the upper part of the soil (McKeague, 1965a). There is little or no mottling (Karavayeva and Targul'yan, 1963). At greater depths in nonpergelic soils, even under conditions of permanent saturation, gleying decreases because of the absence of organic matter and lower biological activity; soil colors are primarily those of the parent materials.

Brown, reddish brown, or yellow mottles form in sections of gleyed profiles that are subject alternately to aerobic and anaerobic conditions. In areas with fairly high rainfall, maximum mottling occurs in the part of the soil column between the highest and lowest levels of the water table (Fig. 27). If the water table never reaches the surface, the soil is not gleyed above the mottled zone. It is commonly gleyed without mottles below that zone, but mottling may occur in the zone of permanent saturation if the groundwater contains dissolved oxygen at times and the parent material is fairly high in iron (McKeague, 1965b). In dry continental areas soils of depressions may be mottled only in the horizons immediately below the 0 horizon as a result of alternate periods of

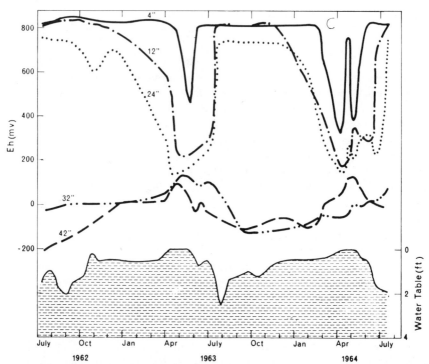

FIGURE 27. *Redox potentials (Eh values) at five depths in a poorly drained soil near Ottawa, Ontario, over a 2-year period. Little or no mottling occurs in horizons with nearly permanent reducing conditions (low or negative Eh values), but mottling is extensive in horizons in which oxidizing and reducing conditions alternate. (Reprinted from McKeague (1965a) by permission of the Agricultural Institute of Canada.)*

dryness and of supersaturation after heavy rains (Karavayeva and Targul'yan, 1963; Rubtsov, 1964). This may be true, also, in upland tundra areas, but in many soils with shallow permafrost tables, maximum mottling is in the part of the active layer that becomes fairly dry in early winter because of out-migration of moisture to both the upper and lower freezing fronts (Pettapiece, 1975).

The mottles that form during aerobic periods consist of fine crystals of insoluble hydrated iron oxides (McKeague *et al.*, 1971) or organo-iron compounds that are resistant to reduction during subsequent anaerobic intervals. Mottling can occur without participation by microorganisms, but in many cases it is the result of the activity of iron-fixing bacteria that accumulate crusts of iron oxide around each cell (Aristovskaya, 1974). In soils that experience large moisture variations in the summer, concretions of organic matter, iron, and manganese may form in the upper part of the soil as a result of biological activity coupled with dehydration and increased concentration of these materials in solution. This is especially true in sandy soils of tundra areas in which there is only a thin zone with periodically aerobic conditions at the surface (Everett, 1979). Once formed, the concretions serve as a nucleus for later crystallization of organometallic compounds and continue to grow (Stenina and Sloboda, 1977).

G. GLEYING IN CALCAREOUS SOILS

Gleying is retarded in strongly calcareous soils with high pH (Drew, 1957; Douglas, 1961). Carbonates do not completely block the process, but have the effect of delaying the onset of reducing conditions in the soil. Also, probably because of the lower availability of iron at high pH, the number of anaerobic iron-reducing bacteria is much lower in calcareous than in acidic soils (Nepomiluyev and Kozyrev, 1970).

H. TRANSLOCATION OF SOLUBLE MATERIALS

Gleyed soils, except for those that are permanently saturated with stagnant water, are subject to leaching and loss of cations and other soluble constituents from the upper horizons. Where the permanent water table is deep, the soluble material may be leached completely out of the soil (Karavayeva, 1973), but more commonly they accumulate in lower horizons and result in an increase in base saturation and pH with depth (Kreida, 1958; Douglas and Tedrow, 1960; Karavayeva, 1963; Allan *et al.,* 1969). Free carbonates may be completely removed from the soil above a permanent water table or permafrost table (Douglas, 1961) or may accumulate as crusts on the lower surface of stones (James, 1970). Leaching effectiveness in poorly drained soils, however, is generally not as great as in associated well-drained soils. Translocation of soluble materials, except by diffusion, is possible only during periods of less than complete saturation when water can percolate through the soil, and leaching is counteracted to a considerable extent by upward movement of large quantities of moisture during freezing. Also, only a small fraction of the annual rainfall actually enters wet soils because of their saturated condition, and lateral movement of moisture is largely confined to the 0 horizon and the upper few centimeters of the mineral soil (Douglas, 1961). The net result in most Cryaquepts is a noticeable but weak leaching regimen.

I. TEXTURE OF GLEYED SOILS

The texture of Cryaquepts is almost entirely dependent on the parent materials of the soils. Any differences in particle-size distribution between horizons that are not directly attributable to heterogeneity of the parent material are invariably slight. Textures range from clay to sand, but the dominant texture of Cryaquepts is silt loam or loam (Douglas and Tedrow, 1960;

Allan *et al.*, 1969; Ignatenko, 1971). This is true in part because of the prevalence of loessial surface deposits in cold areas and in part because the silt-size fraction represents the lower limit of comminution by frost weathering and grinding processes within soils subject to cryoturbation (Karavayeva, 1963). Finer soils occur principally in portions of floodplains far from stream channels that are covered only occasionally by slow-moving water, in beds of former glacial lakes, and in areas of fine-grained shales or other clayey parent materials. Because no cambic horizon can be recognized in sandy or stratified materials, Cryaquepts can be recognized in soils with those textural characteristics only if they have perennially frozen substrata or, if there is no permafrost, an umbric epipedon. Most such soils occur in floodplains or coastal plains.

Because of the low leaching effectiveness in soils that are saturated most of the thaw period, no oriented clay or other evidences of argilluviation are observed in Cryaquepts (Rubtsov, 1964; Pettapiece, 1974). Nevertheless, it is commonly observed that in poorly drained soils underlain by permafrost but not subject to intensive cryoturbation, the clay content in the lower part of the active layer is slightly higher than in the upper horizons, although this is by no means true of all such soils. Reasons for the increased clay at depth are not obvious. The phenomenon has been variously ascribed to weak eluviation (Kreida, 1958; Ignatenko, 1966; Allan *et al.*, 1969), more intense weathering and greater clay formation at depth (Ignatenko, 1963), and loss of clay from upper horizons by selective erosion (Ivanov, 1976).

J. STRUCTURE OF GLEYED SOILS

Soil structure in Cryaquepts, as in most cold soils, is weak. Most of them have no readily apparent structural development, but soils in which the moisture content occasionally drops below saturation develop a granular structure in those horizons that are high in organic matter (Ivanov, 1976) and have platy structure caused by ice lensing in the lower horizons (Drew, 1957; Douglas, 1961; Ugolini, 1966; Pettapiece, 1975). Some soils with permafrost have a coarse, angular blocky structure or breakage in addition to the fine horizontal plates (Douglas and Tedrow, 1960) and vertical cleavage lines that imply the existence of recurring cracks (Ugolini, 1966). The surface horizons of barren spots have the vesicular structure and low bulk density that result from needle-ice development.

K. CLAY MINERALS

In Cryaquepts of the arctic, where chemical weathering is slow, clay minerals in the soil are largely those inherited from the parent material (Ignatenko, 1966). Illite appears to be dominant in most of these very cold soils (Douglas, 1961), but smectites and kaolinite also occur, possibly as a result of synthesis in place from soluble compounds of silicon and aluminum that are trapped above the permafrost. This synthesis probably can take place, however, only as layer additions to previously existing clays. If the soil solution is also high in magnesium, as is likely in lower parts of the landscape where solutes can accumulate, montmorillonite is synthesized. Where the magnesium concentration is low, any newly formed clay is mostly kaolinite (Allan, 1969).

Forested and associated shrub- and sedge-covered Cryaquepts of the boreal belt are fairly high in montmorillonite (Rubtsov, 1964), chloritized clays, and kaolinite (Allan *et al.,* 1969; Pettapiece, 1974). This is especially true in soils that are strongly acidic. Chloritization is apparently related to repeated freezing and thawing, which may have the same effect as frequent wetting and drying in warmer climates (Allan *et al.,* 1969).

L. APPARENT THIXOTROPY

Many gleyed soils of cold areas, particularly those underlain by permafrost, tend to flow after physical agitation and to set when agitation ceases. This is a property akin to thixotropy. Unlike soils high in allophane and allophane-like clays, however, the apparent thixotropy exists only when the soil is saturated and disappears on partial drying (Price *et al.,* 1974). It is most intense in the most highly gleyed horizons in the profile and is absent in any horizons of the same profile that are not gleyed (Rode and Sokolov, 1960; Ignatenko, 1963; Archegova, 1974). It is common in acidic gleyed soils of tundra areas with maritime climates but is seldom observed in soils of areas with strongly continental climates (Karavayeva and Targul'yan 1963) or in calcareous soils (Allan, 1969).

In coarse-silty soils with low clay content the thixotropic-like property probably is due largely to moisture held by capillary forces and to relatively thick water films that coat the silt particles. Agitation or pressure releases the adsorbed water and liquefies the soil (Ignatenko, 1971). In soils with an appreciable clay content the clay tends to form fine platy aggregates under saturated conditions (Liverovskiy *et al.,* 1979), and silicic acid, amorphous aluminum and iron hydroxides, and organometallic complexes are deposited as films on the clay particles. At low temperatures these colloidal films take up large quantities of water. On agitation enough water is released to exceed the upper plastic limit (liquid limit) of the soil, and the soil flows (Allan, 1969).

M. CRYOTURBATION

Cryaquepts, because of their high moisture content, are more subject to cryoturbation than well-drained soils. Cryoturbation is greatest in soils with

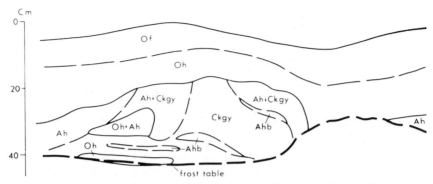

FIGURE 28. *Effect of cryoturbation in a Histic Pergelic Cryaquept of northwestern Canada. 0 horizons are entirely organic; A horizons and inclusions consist of mixed mineral and organic material; and C horizons are predominantly mineral. (From Pettapiece [1975] by permission of the Arctic Institute of North America.)*

shallow ice-rich permafrost in the polar and alpine belt, and most such soils exhibit surface and internal features that are related to frost action (Fig. 28). Not all Cryaquepts with shallow permafrost tables are subject to intensive churning, however. Shrub- and tussock sedge-covered Cryaquepts associated with well-drained forested soils in the boreal belt normally have thick organic mats that permit only very gradual soil temperature and moisture variation in both winter and summer. This reduces and may even prevent the turbation that accompanies more rapid changes in temperature and greater moisture migration within the soil. As an example, a buried organic layer well above the permafrost in such a Cryaquept of interior Alaska has remained essentially undisturbed for several hundred years (Allan *et al.,* 1969). Polygons and other surface expressions of cryoturbation generally are absent in such soils.

II. Classification of Cryaquepts

As a general rule Cryaquepts are poorly drained cold soils in which one or more diagnostic horizons, exclusive of those such as a spodic or argillic horizon that would necessitate classification in an order other than the Inceptisols, can be recognized. There are, however, exceptions to this rule. Soils in which the only diagnostic horizon is a histic epipedon composed entirely of organic soil material are classified as Cryaquents (similar soils with a histic epipedon that contains enough mineral material to remove the horizon from the category of organic soil material remain as Cryaquepts). Poorly drained soils with permafrost are considered to be Cryaquepts even if no cambic horizon can be recognized because of sandy texture, stratification, or a perennially reduced condition. The principal formative process in all of these soils is gleization.

The classification of Cryaquepts at the subgroup level in the United States Taxonomy is fairly complicated and excessively detailed. Simplification, although it would reduce parallelism with the classification of warmer Aquepts, would also reduce the number of meaningless distinctions in the soils of cold areas and would result in better groupings of similar soils. Also, the distinction between Cryaquepts and Cryaquents, based on differentiae at the soil order level, do not always reflect significant differences in soil-forming processes. Combination of these two very similar groups, based on their common aquic moisture regime and the absence of horizons that are diagnostic for other orders, would do much to alleviate complexity in the classification of poorly drained soils.

A. CRYAQUEPTS WITH NO PERMAFROST

Typic Cryaquepts are defined as soils with a gleyed cambic horizon that is thick enough to include all of the soil between depths of 15 and 50 cm, and with no histic or umbric epipedon. They have a mean annual temperature higher than 0° C and contain no layer of volcanic ash as much as 18 cm thick.

Typic Cryaquepts are considerably less extensive than other subgroups of Cryaquepts. They are confined to areas with relatively mild winters where permafrost is absent or rare. Associated well-drained soils on uplands are mostly Spodosols.

Aeric Cryaquepts have only weak gleying, as evidenced by chromas higher than 2 on the Munsell scale in more than 40% of at least one subhorizon between depths of 15 and 50 cm. If the chroma of the soil matrix in all horizons between these depths is higher than 2, regardless of mottling, the soil is not classified as a Cryaquept. *Aeric Humic Cryaquepts* have an umbric epipedon in addition to the weak expression of the gleying process. Soils of both of these subgroups occupy high terraces or foot slopes of forested hillsides in cold continental areas, below steeper slopes on which Cryorthents, Cryochrepts, or Cryorthods are dominant. They commonly have silty or loamy textures. Under the natural vegetation they remain frozen well into the summer, but internal drainage is rapid when the frost disappears. Many exhibit irregular patches and streaks of organic matter throughout the profile, and some are underlain at great depth by blocks of frozen soil, relicts of permafrost developed during an earlier and colder era that were buried under the colluvium that is the parent material of most of these soils. On removal of the forest cover by fire or clearing, these blocks may, after several years, melt and cause pitting or uneven subsidence (Rieger *et al.,* 1963).

Humic Cryaquepts are strongly gleyed throughout the profile and have an umbric epipedon. *Histic Cryaquepts* have a histic epipedon or both histic and umbric epipedons. These soils occur in the same permafrost-free areas as the Typic

Cryaquepts but are normally in even more poorly drained positions. The soil beneath the epipedon may have any texture (except, as noted above, for sandy soils with only a histic epipedon made up of organic material). Humic and Histic Cryaquepts generally exist in close, complex association and have most properties in common.

Andic Cryaquepts have a layer of pyroclastic material in the upper part of the soil that is at least 18 cm thick. They may be formed entirely in volcanic deposits. The soils normally have a thick umbric or histic epipedon. They occupy depressions or seep slopes in areas dominated by Cryandepts and, except for their wetness, have properties like those of the Andepts. In places they are associated with Histosols in depressions.

B. CRYAQUEPTS WITH PERMAFROST

The most extensive subgroups of Cryaquepts are those that are underlain by permafrost. They are by far the dominant soils of the arctic tundra and are also widespread in much of the boreal belt and in high mountains. The soils may have any texture. They support vegetation ranging from sedge or shrub tundra to black spruce, cedar, and other water-tolerant trees. In forested areas, especially under a continental climate, Cryaquepts with permafrost occupy depressions, polar-facing slopes, and seepage areas adjoining well-drained Cryochrepts, Cryumbrepts, or Cryorthods. In parts of the boreal belt with intermittent permafrost, perennially frozen substrata occur only under Cryaquepts and Histosols; the associated well-drained soils are permafrost-free.

Pergelic Cryaquepts have mean annual temperatures of 0° C or less and ice-rich permafrost at some depth. As a rule these soils, because of their thin 0 horizons, thaw to greater depths under comparable climatic conditions than pergelic soils with histic epipedons. They occur, generally, in gravelly or coarse-textured materials and in positions with fairly good surface drainage. They are also extensive in very cold areas where plant growth is so slow that there is little accumulation of organic materials at the surface. Most are strongly mottled and have the distorted textural layers and streaks of organic matter at depth that are characteristic of frost-stirred soils (Fig. 29).

Histic Pergelic Cryaquepts have a histic epipedon and may also have an umbric epipedon above the gleyed subsoil. This is the most extensive of all subgroups of Cryaquepts, in both tundra and forest regions. In most cases the histic epipedon is composed entirely of organic materials in various stages of decomposition. Because of the thick accumulation at the surface, the soils thaw to only shallow depths during the summer. In some exceptionally cold years, thawing may be only in the histic epipedon and may not reach the underlying mineral soil. The mineral soil may be mottled or, in the wettest positions, may have dull colors without mottling. Depending on the depth of thaw, cryoturbation may

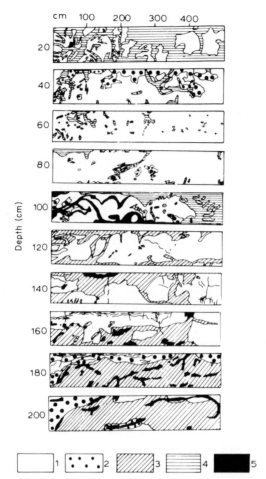

FIGURE 29. *Distribution of organic and mineral matter in sections of a Pergelic Cryaquept with a deep permafrost table: (1) yellowish brown mineral soil; (2) mottled mineral soil; (3) mineral soil with some organic mixture; (4) well-decomposed organic matter; (5) concentrated, coal-like organic matter. (From O. W. Makeev and A. S. Kerzhentsev [1974]. Cryogenic processes in the soils of northern Asia.* Geoderma 12, *101–109.)*

be severe or may be nearly absent. Where extensive cryoturbation has taken place, there is commonly an accumulation of organic matter at the permafrost table.

Pergelic Ruptic–Histic Cryaquepts have a histic epipedon in only part of each pedon. They form where cryoturbation has forced mineral soil to the surface and disrupted the organic mat. This takes place most commonly in fine-silty or clayey soils in areas where the growth of ice wedges in the permafrost has created a polygonal surface pattern. In most pedons of this subgroup, the histic epipedon is confined to the outer portions of the polygon and the troughs between polygons. The center of the polygon is barren or has a sparse vegetative cover made up largely of lichens and other pioneer species. In gravelly material pebbles and stones are more numerous at the surface of the barren spot, or frost circle, than in the whole soil or, because the surface is normally slightly convex, they may form a garland at the perimeter of the spot. In some soils greater evaporation at the surface in summer and more rapid freezing in early winter has resulted in a thin crust of soluble salts at the surface of the barren areas (Ignatenko, 1967b). This is partially reversed in the spring by migration of soluble constituents from the soil under the frost circle, which thaws earlier, to the still-frozen soil under vegetation (Vasil'yevskaya, 1979). The permafrost table is deeper under the circle than under the vegetated portions of the pedon.

Barren spots also occur commonly on solifluction slopes, where they are formed in the upslope portions of lobes as a result of movement of the soil mass below them. In this situation, however, only isolated pedons of Pergelic Ruptic–Histic Cryaquepts occur; most pedons in the lobes are fully vegetated.

Humic Pergelic Cryaquepts have an umbric epipedon rather than a histic epipedon above the gleyed horizon. They occupy depressions and slopes subject to seepage in high mountains at mid latitudes. They also occur in dunes in the arctic where decomposition of organic matter in soils of slightly elevated sites and additions of blown sand reduce the organic matter percentages in the 0 horizon to the point that a histic epipedon can no longer be recognized (Everett, 1979). The permafrost table in these soils is generally deep.

C. OTHER SUBGROUPS

Lithic Cryaquepts overlie bedrock at a depth of less than 50 cm below the surface of the mineral soil.* They may have temperatures either above or below 0° C. *Histic Lithic Cryaquepts* have a histic epipedon above the gleyed horizon. Lithic Cryaquepts occur in areas of very thin glacial till over scoured bedrock and on weathered rock outcrops in mountains. Histic Lithic Cryaquepts are, mostly, on steep, polar-facing slopes in nonglaciated cold areas. They may have a perennially frozen layer above the bedrock. Both subgroups are inextensive.

Subgroups of *Halic Cryaquepts* and *Halic Pergelic Cryaquepts,* although not presently recognized, are needed for poorly drained soils formed in saline materials in which there has been no translocation of clay. Definitions of the subgroups should be comparable, except for temperature restrictions, to that for the Typic Halaquepts of temperate regions. Halaquepts, apparently without permafrost, have been described in northwestern Canada (Pringle *et*

---

* This subgroup is not listed in the United States Taxonomy, although the soils are known to exist in Alaska (Rieger *et al.,* 1979).

al., 1975) and in the vicinity of hot springs in Alaska (Krause and Wilde, 1966). They occur widely over permafrost in dry continental areas of eastern Siberia (Yelovskaya, 1965; Konorovskiy, 1976), especially in depressions in floodplains and terraces. Poorly drained saline soils also occur in the high arctic in recently elevated marine sediments (Smith, 1956) and in other materials where salts are added to the soils by ocean spray (Svatkov, 1958).

References

Allan, R. J. (1969). "Clay Mineralogy and Geochemistry of Soils and Sediments with Permafrost in Interior Alaska." Unpub. Ph.D. Dissertation, Dartmouth C.

Allan, R. J., Brown, J., and Rieger, S. (1969). Poorly drained soils with permafrost in interior Alaska. *Soil Sci. Soc. Am. Proc.* **33**, 599–605.

Archegova, I. B. (1974). Humus profiles of some taiga and tundra soils in the European U. S. S. R. *Sov. Soil Sci.* **6**, 136–141.

Aristovskaya, T. V. (1974). Role of microorganisms in iron mobilization and stabilization in soils. *Geoderma* **12**, 145–150.

Bel'chikova, N. P. (1966). Characteristics of humic substances of taiga soils in central Siberia which have developed on basic rocks. *Sov. Soil Sci.*, 1136–1145.

Billings, W. D., Peterson, K. M., Shaver, G. R., and Trent, A. W. (1977). Root growth, respiration, and carbon dioxide evolution in an arctic tundra soil. *Arc. Alp. Res.* **9**, 129–137.

Boyd, W. L., and Boyd, J. W. (1971). Studies of soil microorganisms, Inuvik, Northwest Territories. *Arctic* **24**, 162–176.

Bunting, B. T., and Hathout, S. A. (1971). Physical characteristics and chemical properties of some high-arctic organic materials from southwest Devon Island, Northwest Territories, Canada. *Soil Sci.* **112**, 107–115.

Cameron, R. E., King, J., and David, C. N. (1970). Soil microbial ecology of Wheeler Valley, Antarctica. *Soil Sci.* **109**, 110–120.

Daragan, A. Yu. (1967). Microbiology of the gley process. *Sov. Soil Sci.*, 228–236.

Douglas, L. A. (1961). "A Pedologic Study of Tundra Soils from Northern Alaska." Unpub. Ph.D. Dissertation, Rutgers U.

Douglas, L. A., and Tedrow, J. C. F. (1960). Tundra soils of arctic Alaska. *Trans. 7th Int. Cong. Soil Sci.*, *Vol. IV*, 291–304.

Drew, J. V. (1957). "A Pedologic Study of Arctic Coastal Plain Soils Near Point Barrow, Alaska." Unpub. Ph.D. Dissertation, Rutgers U.

Everett, K. R. (1979). Evolution of the soil landscape in the sand region of the arctic coastal plain as exemplified at Atkasook, Alaska. *Arctic* **32**, 207–223.

Everett, K. R., and Parkinson, R. J. (1977). Soil and landform association, Prudhoe Bay area, Alaska. *Arc. Alp. Res.* **9**, 1–19.

Holowaychuk, N., Petro, J. H., Finney, H. R., Farnham, R. S., and Gersper, P. L. (1966). Soils of Ogotoruk Creek watershed. *In* "Environment of the Cape Thompson Region, Alaska" (N. J. Wilimovsky, ed.), pp. 221–273. U. S. Atomic Energy Comm., Div. Tech. Inf. Ext., Oak Ridge, Tennessee.

Ignatenko, I. V. (1963). Arctic tundra soils of the Yugor Peninsula. *Sov. Soil Sci.*, 429–440.

Ignatenko, I. V. (1966). Vaygach Island soils. *Sov. Soil Sci.*, 991–998.

Ignatenko, I. V. (1967). Waterlogged and bog soils of the East European tundra. *Sov. Soil Sci.*, 177–192. (a)

Ignatenko, I. V. (1967). Soil complexes of Vaygach Island. *Sov. Soil Sci.,* 1216–1229. (b)

Ignatenko, I. V. (1971). Soils in the Kara River basin and their zonal position. *Sov. Soil Sci.* **3,** 15–28.

Ivanov, V. V. (1976). Possible role of surficial soil-eluvial processes in the genesis of the mantle deposits of the north. *Sov. Soil Sci.* **8,** 651–657.

Ivanova, Ye. N. (1965). Frozen taiga soils of northern Yakutia. *Sov. Soil Sci.,* 733–744.

Ivarson, K. C. (1965). The microbiology of some permafrost soils in the Mackenzie Valley, N. W. T. *Arctic* **18,** 256–260.

James, P. A. (1970). The soils of the Rankin Inlet area, Keewatin, N. W. T., Canada. *Arc. Alp. Res.* **2,** 293–302.

Jordan, D. C., Marshall, M. R., and McNicol, P. J. (1978). Microbiological features of terrestrial sites on the Devon Island lowland, Canadian Arctic. *Can. J. Soil Sci.* **58,** 113–118.

Karavayeva, N. A. (1963). Description of arctic-tundra soils on Bol'shoi Lyakhovskii I. (Novosibirskie Islands). *In* "Soils of Eastern Siberia" (Ye. N. Ivanova, ed.), pp. 123–146. U. S. Dept. Comm., Clearinghouse for Fed. Sci. and Tech. Inf., Trans. TT 69-55073, Springfield, Virginia.

Karavayeva, N. A. (1973). Acid eluvial-gley soils in the middle and northern taiga of West Siberia. *Sov. Soil Sci.* **5,** 129–143.

Karavayeva, N. A. (1974). Major kinds of gley soils of the tundra and northern taiga regions in the Soviet Union. *Geoderma* **12,** 91–99.

Karavayeva, N. A., and Targul'yan, V. O. (1960). Humus distribution in the tundra soils of northern Yakutia. *Sov. Soil Sci.,* 1293–1300.

Karavayeva, N. A., and Targul'yan, V. O. (1963). Contribution to the study of soils on the tundras of northern Yakutia. *In* "Soils of Eastern Siberia" (Ye. N. Ivanova, ed.), pp. 57–78. U.S. Dept. Comm., Clearinghouse for Fed. Sci. and Tech. Inf., Trans. TT 69-55073, Springfield, Virginia.

Konorovskiy, A. K. (1976). Characteristics of formation and properties of the floodplain soils of central Yakutiya. *Sov. Soil Sci.* **8,** 125–132.

Krause, H. H., and Wilde, S. A. (1966). Solonchak soils of Alaska. *Sov. Soil Sci.,* 43–44.

Kreida, N. A. (1958). Soils of the eastern European tundras. *Sov. Soil Sci.,* 51–56.

Leahey, A. (1947). Characteristics of soils adjacent to the Mackenzie River in the Northwest Territories of Canada. *Soil Sci. Soc. Am. Proc.* **12,** 458–461.

Liverovskiy, Yu. A., Zverena, T. S., and Il'inskaya, G. G. (1979). Characteristics of the genesis of the tundra soils of the northern Ob' region. *Sov. Soil Sci.* **11,** 512–520.

Makeev, O. W., and Kerzhentsev, A. S. (1974). Cryogenic processes in the soils of northern Asia. *Geoderma* **12,** 101–109.

McKeague, J. A. (1965). Properties and genesis of three members of the Uplands catena. *Can. J. Soil Sci.* **45,** 63–77. (a)

McKeague, J. A. (1965). A laboratory study of gleying. *Can. J. Soil Sci.* **45,** 199–206. (b)

McKeague, J. A., Brydon, J. E., and Miles, N. M. (1971). Differentiation of forms of extractable iron and aluminum in soils. *Soil Sci. Soc. Am. Proc.* **35,** 33–38.

Moore, T. R. (1973). The distribution of iron, manganese, and aluminum in some soils from north-east Scotland. *J. Soil Sci.* **24,** 162–171.

Nepomiluyev, V. R., and Kozyrev, M. A. (1970). Gleying in soil formation and participation of microorganisms therein. *Sov. Soil Sci.* **2,** 560–565.

Pettapiece, W. W. (1974). A hummocky permafrost soil from the subarctic of northwestern Canada and some influences of fire. *Can. J. Soil Sci.* **54,** 343–355.

Pettapiece, W. W. (1975). Soils of the subarctic in the lower Mackenzie basin. *Arctic* **28,** 35–53.

Price, N. W., Bliss, N. C., and Svoboda, J. (1974). Origin and significance of wet spots on scraped surfaces in the high arctic. *Arctic* **27,** 304–306.

Pringle, W. L., Cairns, R. R., Hennig, A. M. F., and Siemens, B. (1975). Salt status of some soils of the Slave River lowlands in Canada's Northwest Territories. *Can. J. Soil Sci.* **55**, 399–406.

Rieger, S., DeMent, J. A., and Sanders, D. (1963). "Soil Survey of Fairbanks Area, Alaska." U. S. Dept. Agr., Soil Cons. Serv., Ser. 1959, No. 25,

Rieger, S., Schoephorster, D. B., and Furbush, C. E. (1979). "Exploratory Soil Survey of Alaska." U. S. Dept. Agr., Soil Cons. Serv. Washington, D. C.

Rode, T. A., and Sokolov, I. A. (1960). The Transbaykal mountain-tundra landscapes. *Sov. Soil Sci.,* 384–391.

Rubtsov, D. M. (1964). Gley weakly-podzolic soils. *Sov. Soil Sci.,* 677–681.

Smith, J. (1956). Some moving soils in Spitsbergen. *J. Soil Sci.* **7**, 10–21.

Stenina, T. A., and Sloboda, A. V. (1977). New humus formations in tundra soils. *Sov. Soil Sci.* **9**, 414–416.

Sushkina, N. N. (1960). Characteristics of the microflora of arctic soils. *Sov. Soil Sci.,* 392–400.

Svatkov, N. M. (1958). Soils of Wrangel Island. *Sov. Soil Sci.,* 80–87.

Tedrow, J. C. F. (1969). Thaw lakes, thaw sinks, and soils in northern Alaska. *Biul. Peryglac jalny* **20**, 337–344.

Ugolini, F. C. (1966). Soils of the Mesters Vig District, northeast Greenland: II. Exclusive of Arctic Brown and Podzol-like soils. *Medd. om Gron., Bd. 176, No. 2,* 25 pp.

Vasil'yevskaya, V. D. (1979). Genetic characteristics of soils in a spotty tundra. *Sov. Soil Sci.* **11**, 390–401.

Yelovskaya, L. G. (1965). Saline soils of Yakutia. *Sov. Soil Sci.,* 355–359.

Chapter *12*

HISTOSOLS

Histosols are soils that consist wholly or dominantly of organic soil materials. In most cases the peat is thick and the organic nature of the soil is easily recognizable, but where it is only moderately thick over a mineral substratum or is interspersed with layers of mineral material, a rather complex definition is needed to distinguish between Histosols and soils of other orders. A soil is considered to be a Histosol if at least one-half (three-quarters in the case of fibrous, usually sphagnum, moss peat), by volume, of the upper 80 cm of the soil consists of organic soil material that is usually saturated during the thaw period. Thus, except in the case of fibrous moss peat, a mineral layer up to 40 cm thick may overlie a peat deposit, mineral layers with a cumulative thickness of as much as 40 cm may occur within the upper 80 cm, or surface peat may overlie mineral material at depths below 40 cm. Fibrous peat must be at least 60 cm thick (i.e., there can be only 20 cm of mineral material in the upper 80 cm of the soil). The reason for the greater thickness requirement for fibrous moss peat is the very low bulk density of that material (normally less than 0.1 g/cm^3) as compared with other peats and its probable greater degree of subsidence after drainage. Also classified as Histosols are soils with less than 40 cm of organic material, whether or not it is usually saturated, if the peat rests directly on gravel or stones with few fines (fragmental material) or if there is less than 10 cm of mineral soil between the base of the peat and bedrock or other

179

consolidated material, the peat is at least twice as thick as that layer of mineral soil, and no horizons diagnostic for other orders have developed.

A soil in which the surface layer is organic material that is not usually saturated, such as the 0 horizon of a well-drained soil, is not classified as a Histosol even if the 0 horizon is thicker than 40 cm unless the 0 horizon rests on fragmental material or bedrock as described above. Soils that consist solely of comparatively dry plant litter over these substrata are included as a separate suborder in the order of Histosols and are the only Histosols that are naturally well-drained.

I. Genesis of Histosols

The accumulation of peat, or the process of *paludification* (Auer, 1930), can take place under water, as in a lake or pond, or in uplands on soils that are nearly constantly wet or that are very slowly permeable. In cool maritime areas, especially at elevations above tree line, thick deposits of peat commonly accumulate even on exposed bedrock. Many different plant species contribute to the peat, but most of it is derived from either mosses, sedges and reeds, or woody plants. The living vegetation on organic soils varies with local conditions of climate, groundwater, and relief. In many cases the vegetation is different in different portions of a single bog, and major changes in vegetation commonly take place as the peat accumulates.

Peat bogs can be conveniently divided, on the basis of position in the landscape and mode of development, into level or lowland bogs, convex or raised bogs, and sloping or blanket bogs (Dachnowski-Stokes, 1941). Of these, the lowland bogs are most widespread.

A. DEVELOPMENT SEQUENCE IN LOWLAND BOGS

Many lowland bogs are the result of gradual filling by organic materials of lakes and ponds. Bogs of this kind are especially prevalent in glaciated areas and in abandoned stream channels and other backwater areas of floodplains and deltas. Some develop in ponds formed by beaver dams in small stream channels. In a typical developmental sequence, limnic materials including remnants of floating plants, fecal pellets and other remains of aquatic animals and microorganisms, and, in some cases, diatoms or the shells of freshwater crustaceans, form the initial deposit on the floor of these quiet water bodies. Sedges and reeds growing in poorly drained soils at the perimeter of the lakes gradually spread inward and form a layer of plant remains over the original deposits. In time these remains completely fill the lake, and the former lake surface, now the water table, lies slightly below the surface of the peat. At this point the peat is colonized by mosses, commonly *sphagnum* species in cold areas, which build up a layer of moss peat above the water table. Although the mosses derive all of their moisture from the atmosphere, the water-holding ability of both the living moss and the peat below it is so great that they are always wet. Water-tolerant trees—black spruce, tamarack, lodgepole pine, cedar, and others—and shrubs may eventually cover the peat surface and contribute a layer of woody fragments.

Despite this general pattern of development, the sequence, in ascending order, of limnic materials, sedge peat, moss peat, and woody peat is seldom completely regular. Any of the sequential layers can be absent or in a different position as a consequence of a change in the environment of a bog. Fluctuations in the water level caused by topographic changes, modifications in stream flow patterns, climatic variations, or wildfires can temporarily alter or even reverse the order of accumulation. Layers of sedge peat, for example, may have formed within or above accumulations of moss peat, and layers of woody peat, formed during periods of warmer or drier environments, can occur below the sedge or moss peat. Deposits of mineral material, such as alluvium or volcanic ash, can modify the sequence for a time. In very cold areas permafrost can form below an accumulation of peat and have a considerable effect on its further development.

Lowland bogs also cover extensive areas with high water tables. One such region is the Hudson Bay Lowland, an area of recent marine transgression about 1300 km long and 300 km wide (Sjörs, 1959). Another is in northwestern Siberia, where about 40% of the land area consists of organic soils with permafrost (Firsova, 1967). Many smaller but similar lowland areas of peat soils that have no history of lake-filling also occur in both the boreal belt and the arctic tundra. In places the peat accumulates over older Spodosols (Page *et al.*, 1980). In most cases the peat forms above mineral soils that are constantly wet because of high groundwater levels, low permeability, or a combination of high rainfall and low rates of evapotranspiration. Paludification also occurs following destruction of a forest by fire or clearing. Because of the reduction in transpiration the level of the water table rises, and this in turn encourages the growth of peat-forming sedges and mosses and the eventual development of an organic soil. No limnic layers occur in these soils (although the underlying soil may be calcareous). Where mineralized, nonacid groundwater moves slowly through the peat that has built up above the wet mineral substratum, the vegetation consists principally of sedges, mosses other than sphagnum, and some stunted trees and shrubs. The peat itself in this situation is nonacidic and fairly high in calcium and other cations (Sjörs, 1959; Zoltai and Tarnocai, 1975). Where rain is the principal source of moisture, the peat is largely derived from sphagnum moss and tends to be strongly acid. Acidic *sphagnum* moss peat also occurs at the borders of pools filled by rainwater within areas of mineralized peat and may make up the surface vegetation on hummocks, peat ridges, and other slightly elevated portions of the bog.

B. HUMMOCK AND PALSA FORMATION

Most lowland bogs have an irregular surface, consisting of hummocks or palsas as much as several meters high that are separated by depressions. The irregular surface is most pronounced in areas within or close to the transition zone between forest and tundra vegetation. In many cases the hummocks have a perennially frozen core, and the intervening portions of the bog are permafrost-free (Ignatenko, 1967; Zoltai and Tarnocai, 1971). The original irregularity probably results from differences in vegetation height. Snow is blown from the high points and accumulates in the lower portions of the bog. As a result, the peat under the high points freezes more deeply in winter and, because of its relative dryness, thaws more slowly in summer. The core of ice grows annually because of migration of water along the thermal gradient thus formed and increases the height of the hummocks (Troll, 1958; Zoltai and Tarnocai, 1971) (Fig. 30). A similar situation exists in peats underlain by con-

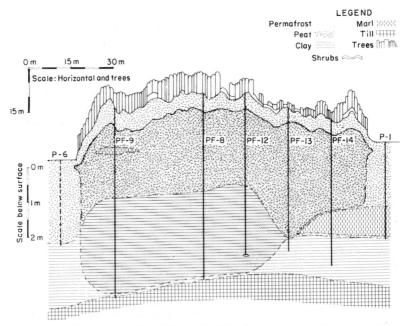

FIGURE 30. *Section through a palsa in northern Manitoba, showing the perennially frozen core. In contrast with the surrounding bog, the elevated and somewhat drier palsa is forested. (From Zoltai and Tarnocai [1971] by permission of the University of Colorado.)*

tinuous permafrost in which ice-wedge polygons have developed. Here the troughs between high-centered polygons commonly have water at the surface throughout the summer. The troughs widen in time, creating a mosaic of sedge peat in low areas and, commonly, *sphagnum* peat on the drier centers. The eventual result is a few isolated mounds with a moss cover in an extensive marsh (Brown, 1966). In peat forming in extremely wet areas such as deltas, the polygon edges are raised, and the centers of the polygons are flooded. In this situation the drier peat is on the raised edges. The pattern is eventually destroyed as the peat continues to accumulate and fill in the centers (Zoltai and Tarnocai, 1975) or, alternatively, as the ponds grow and coalesce to form lakes.

Extreme examples of the development of palsas are forested "peat plateaus," which are higher than the surrounding treeless bog and which may cover areas larger than 1 km². These plateaus form, normally, in the boreal belt close to the fringe of the area of continuous permafrost beneath dense stands of black spruce. Because of lower soil temperatures under the trees and drier soil conditions as a result of increased transpiration, a layer of permafrost develops, and the soil surface is elevated as ice lenses grow. In contrast with the peat in the unfrozen lower portions of the bog, which is derived principally from sedges, a surface layer of *sphagnum* moss peat forms under the trees. The peat plateaus are fairly unstable. The edges of the pleateau tend to slough off into the surrounding bog, and removal of the trees, as by fire, can cause the plateau to collapse in whole or in part as the permafrost melts. This can result in a complex pattern of frozen and unfrozen peat (Sjörs, 1959; Zoltai and Tarnocai, 1971; Tarnocai, 1972). Comparable "spruce islands" with no permafrost also exist in lowland bogs in the warmer part of the boreal belt. In those areas elevations of the "islands" above the general surface of the bog are not as great, and the surrounding peat also has sphagnic surface layers. Drier surface conditions under the trees are apparently due entirely to greater transpiration.

C. STRING BOGS

Where a bog has a slight slope and there is seepage through it in a consistent direction, a pattern of sinuous ridges or "strings" roughly perpendicular to the direction of the flow develops (Troll, 1958). This takes place in bogs both with and without permafrost. During the spring thaw plant debris carried by moving water over the still frozen peat catches on higher vegetation, particularly at the edges of temporary cross-channels between thaw pools. When the snow melt is completed and the flow of water abates, the debris lines become preferred sites for shrubs because of their relatively drier condition. Eventually they attain sufficient elevation to act as permanent dams. During the freezing period ice forming in the wetter peat adjacent to the debris lines elevates them further, narrows them, and prevents their migration in the direction of water flow. Debris lines develop first at the downstream end of a bog, then form progressively closer to the upstream edge (Thom, 1972).

The final step in the development of bogs formed as a result of the filling of relatively small lakes or depressions is a convex or raised bog. After the filling sequence has progressed to the point that *sphagnum* moss peat makes up the upper layer of the entire bog, the growth rate of the moss tends to be greater in the center than at the edges of the bog. This may be due in part to less shading by the surrounding forest and in part to somewhat drier conditions in positions less subject to inundation by runoff from adjoining land. As a result of the differential in the rate of growth, the bog assumes the form of a dome in which the center may eventually be as much as several meters higher than the edges (Fig. 31). Shallow pools of open water are common in the high central portion, possibly the result of local concentrations of water from thawing snow and ice. In the later stages of development, a variety of shrubs and trees such as black spruce and lodgepole pine colonize the peat (Dachnowski-Stokes, 1941).

Convex or raised bogs can occur in any climate, but they have their maximum development under maritime climates with high humidity throughout the year. In such a climate rainwater running off the slopes of the dome increases the wetness of adjoining soils and can result in expansion of the bog into surrounding forested land.

Sloping bogs are most common in areas with cool maritime climates such as the northern British Isles (Proudfoot, 1958; Romans *et al.*, 1966), New-

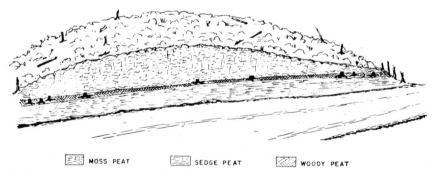

MOSS PEAT   SEDGE PEAT   WOODY PEAT

FIGURE 31. *Section through a convex bog in southeastern Alaska, with sedge peat overlain by* sphagnum *moss peat in the shape of a dome. A coniferous forest existed at the site for a time and was later succeeded by the moss. (From Dachnowski-Stokes, 1941.)*

foundland (Brewer *et al.*, 1973), the Aleutian Islands (Everett, 1971), and areas bordering the Gulf of Alaska (Rieger *et al.*, 1979), where they occupy a significant portion of the land surface. They also occur in high mountains at lower latitudes (Retzer, 1956) and on steep, polar-facing slopes in more continental climates, where the lower layers of the peat may be perennially frozen.

Under maritime climates sloping organic soils form where the underlying material is impermeable or only slowly permeable and where there is a nearly constant flow of seepage water or very high precipitation rates throughout the year. These conditions are met most commonly in areas of compact glacial till at lower elevations, on the lower slopes of steep hills, and in areas of exposed bedrock or glacial till at elevations above tree line. Typically, sloping bogs are made up dominantly of more or less fibrous or "felty" sedge peat arranged in layers parallel with the underlying mineral surface. In some cases buried woody fragments indicate that the sedge vegetation has replaced a previously existing forest; some sloping organic soils have layers of woody peat that represent intervals of relative dryness during which forests were reestablished on the peat substrata (Fig. 32). The upper layers of the peat or, less commonly, the entire bog, may contain an admixture of *sphagnum* fibers or may consist entirely of moss peat. In many places, where enough peat has accumulated so that the groundwater level no longer reaches the surface, water-tolerant coniferous trees (cedar, hemlock, lodgepole pine), shrubs, and herbaceous plants colonize the bog.

Peat on slopes that are only intermittently saturated but in which the water

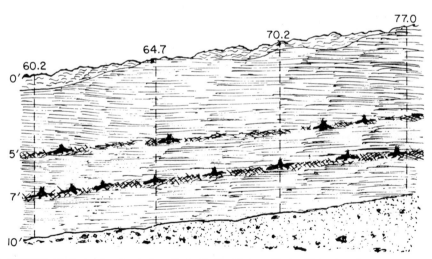

FIGURE 32. *Section through a sloping or blanket bog in southeastern Alaska. The bog is made up principally of sedge peat except for an intermittent upper layer of* sphagnum *moss peat. Layers of woody peat formed during relatively dry intervals when forests were established on the peat. (From Dachnowski-Stokes, 1941.)*

table is usually high is commonly highly decomposed and only moderately thick over a mineral substratum. Spodosol profiles may underlie the peat (Proudfoot, 1958; Romans *et al.,* 1966; McKeague *et al.,* 1968). Most organic soils of this kind are forested, but have a high proportion of succulent plants in the ground cover. Relatively thin accumulations of highly decomposed peat also occur on slopes, including sloping bedrock outcrops, above tree line in very humid climates (Everett, 1971; Rieger *et al.,* 1979). It is likely that, if wet conditions persist, sedges will eventually replace the existing vegetation on many of these soils and accumulation of sedge peat will begin.

F. PERMAFROST

In the central part of the boreal belt permafrost can develop in organic soils as a consequence of the accumulation of peat, even in areas where adjacent mineral soils do not have perennially frozen substrata. In the arctic and in the colder boreal areas, permafrost is present at some depth from the inception of peat accumulation. Such soils commonly have considerable intermixing of organic and mineral materials at the base of the peat. In either case the permafrost table rises gradually as the peat accumulates and, barring disturbance of the vegetation or upper horizons, eventually is well above the interface between organic and mineral soil.

Organic soils with permafrost may occupy any position, from lowland to strongly sloping, and may consist of woody, sedge, or moss peat. In most cases, especially in the zone of continuous permafrost, the surface of the peat is irregular as a result of freezing processes.

G. PLACIC HORIZONS

Placic horizons, some in association with a spodic horizon, commonly occur in the mineral soil below "blanket bogs" on rolling topography. In coarse-grained materials there may be a series of these horizons in and below a spodic horizon (McKeague *et al.,* 1968), but in most cases single, thin placic horizons occur immediately below the interface between the peat and the mineral soil (Fig. 33). The placic horizons under peat differ from those that occur in well-drained soils in that they contain little or no organic matter. They consist of two distinct parts: the upper one, ferruginous and highly indurated; and the lower one, black, high in manganese, and less strongly cemented (McKeague *et al.,* 1968). Both the iron and the manganese probably are transported to the bottom of the peat in the reduced divalent state, but the mechanism of oxidation and precipitation in the placic horizon is unclear. It may be the result of microbial activity at the base of the peat (Brewer *et al.,* 1973) or it may be accomplished without biological intervention (McKeague *et al.,* 1968). In either event, it probably takes place under relatively dry conditions that exist in midsummer

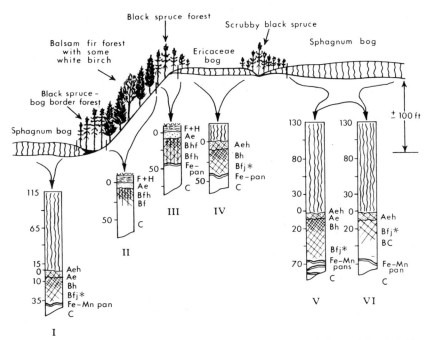

FIGURE 33. *Position of placic horizons (Fe pans and Fe–Mn pans) in some soils of outwash deposits in New-foundland. These placic horizons, which occur below thick peat deposits and spodic horizons beneath the peat, differ from placic horizons in well-drained Spodosols in that no migration of organometallic complexes appears to be involved. (From McKeague et al. [1968] by permission of the Agricultural Institute of Canada.)*

between the peat and the permanent groundwater table, especially where the underlying material is compact and nearly impervious. In some areas the subpeat soil can be dry even as water runs into ditches from the base of the peat (Proudfoot, 1958). In soils with albic and spodic horizons under the peat, the placic horizons may have formed before the onset of peat accumulation. This is indicated in one such soil in Scotland, in which the pollen assemblage below the placic horizon is of an earlier age than that above the horizon (Romans *et al.*, 1966).

H. OTHER IRON DEPOSITS

Iron moving through organic soils in the ferrous form or in combination with organic acids is commonly precipitated in more diffuse form than a placic horizon. It may appear as stains on fracture planes of finely divided peat or in a zone at the contact of the peat with the underlying mineral soil (Everett, 1971), or as mottles below the peat. If the iron comes into contact with groundwater that contains dissolved oxygen, it may precipitate as bog iron either below the peat or in an adjoining drainage basin (Kovalev and Generalova, 1967). Bog iron differs from placic horizons in that it forms in thicker sheets and does not exhibit their extreme induration or distinct layering.

II. Properties of Histosols

A. CLASSIFICATION OF ORGANIC SOIL MATERIALS

In assessing the suitability of organic soils for agriculture and other uses, degree of decomposition of the peat is in many ways a more appropriate means for distinguishing between peats than botanic origin. Three kinds of peat have been defined on this basis for use in the United States Taxonomy. *Fibric soil materials* are no more than slightly decomposed. Even with rubbing, most of the peat resists complete breakdown of fibers, and the plants from which they are derived can still be identified with the aid of a hand lens. If fibers do break down mechanically under vigorous rubbing, very little of the organic matter can be dissolved in a solution of sodium pyrophosphate.* Most fibric material is derived from mosses and woody plants, but in cold areas especially, fibrous sedge peat also exists. *Hemic soil materials* are partially decomposed. Most of the fibers are destroyed by rubbing, and more organic matter is dissolved in the pyrophosphate extract than in the case of fibric materials. Hemic material in cold areas is derived mostly from sedges and reeds. *Sapric soil materials* are highly decomposed, have few or no remaining fibers, and are highly soluble in pyrophosphate. The materials are derived mainly from broad-leaved plants in cold areas, but some is formed from sedge and even woody plant remains. Most sapric peats contain a higher proportion of mineral matter than fibric peats, which generally consist exclusively of organic matter except in cases where mineral material such as alluvium or volcanic ash occurs in layers in the peat (Walmsley and Lavkulich, 1975).

Other materials, such as coprogenous earth made up largely of fecal pellets from aquatic animals, diatomaceous earth from fossils of silica-rich algae, and marl from the shells of freshwater crustaceans, also occur in or below organic soils. These *limnic materials,* if present, can be significant in the utilization of organic soils, but they are presently recognized in classification only at the subgroup and lower categories.

B. GENERAL PROPERTIES

The physical properties of peat are related to its fiber content and degree of decomposition. Fibric peat has a dry bulk density in the range of .05–.1 g/cm³, as opposed to bulk densities of .25–.5 g/cm³ in sapric peats. The values for

---

* Quantitative criteria for fiber content after rubbing and degree of solubility as determined by the color of the pyrophosphate extract are given in *Soil Taxonomy* (Soil Survey Staff, 1975).

hemic peats, as for most properties, are intermediate. Most organic soils have the ability to adsorb quantities of water well in excess of their dry weight, but fibric peats release the moisture more readily and have greater hydraulic conductivity or permeability than the others. Most fibric peat soils, therefore, are poorer storage reservoirs for water than loamy mineral soils. Organic soils, especially sapric peats, have poor capacity for reabsorbing moisture after air drying (Bukhman, 1960; Farnham and Finney, 1965; Tarnocai, 1972; Walmsley and Lavkulich, 1975; Silc and Stanek, 1977).

The ice content of permafrost in organic soils is exceptionally high, ranging from 60% to 90% on a volume basis (Fig. 34). As a result they are even more vulnerable than mineral soils to pitting and other forms of degradation when the vegetation or the upper layers of the soil are disturbed (Tarnocai and Zoltai, 1978).

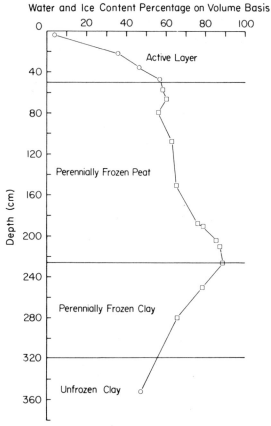

FIGURE 34. *The ice content of a perennially frozen Histosol in northern Manitoba. (From Tarnocai [1972] by permission of the Agricultural Institute of Canada.)*

Fibrous *sphagnum peat* is invariably extremely acid and low in exchangeable cations (Bukhman, 1960). Most other cold organic soils, except for those that form in areas where the groundwater is high in bases, are also acidic and nutrient-poor, although not to the same extent as *sphagnum* peat. The carbon to nitrogen ratio is typically very wide in fibric peats but is in essentially the same range as mineral soils in sapric peats.

Strongly acidic fibric peats have low levels of microbial activity throughout, but most other peats have an upper layer with a large population of microorganisms and a lower layer that is relatively inert. The microbes are especially abundant in any zone in the upper part of the soil that is occasionally aerated by a drop in the groundwater level. It is in such a zone that the greatest amount of decomposition and conversion of fibric to hemic and sapric peat takes place. Where no zone of concentration of living organisms exists, decomposition rates are slow (Ingram, 1978).

III. Classification of Cold Histosols

The classification of Histosols in the United States Taxonomy is centered on the upper 130 cm of the profile (160 cm if the upper 60 cm consists of fibric peat with a bulk density of less than .1 g/cm^3), even if the lower part of this "control section" is mineral rather than organic material. It differs from the classification of mineral soils in that the concept of a control section applies at all levels of the taxonomic scheme within the order rather than just in the lowest categories. The control section is shortened in only three situations. If bedrock or other consolidated material occurs within the 130 cm (or 160 cm) depth, it is taken as the base of the section. If permafrost occurs within these depths, the base of the control section is 25 cm below the permafrost table. In the case of peat that actually is floating on free water, the control section extends only as far as the water surface.

The control section is divided into three tiers—surface, subsurface, and bottom. The surface tier is the upper 30 cm of the soil except where fibric peat with low bulk density makes up the upper 60 cm; in that case the surface tier is 60 cm thick. The subsurface tier begins at the base of the surface tier and is 60 cm thick. The bottom tier is 40 cm thick. In soils with shortened control sections the tier thicknesses are the same, but a portion of the surface tier or a part or all of the lower tiers may be absent.

The reliance on arbitrary depths in the classification of organic soils, as opposed to recognition of natural distinctions such as a division into biologically active and inert zones, has been severely criticized (Ingram, 1978). The variability in bog formation that results from changing external conditions that differ from site to site, however, and the difficulty in identifying natural zones in the field make the biological approach to classification impractical. General trends can be discerned and the history of each bog can be determined from the sequence of layers within it, but local differences are so numerous that some degree of rigidity, at least with respect to depth criteria, seems to be necessary in a general classification scheme.

For all except the thin soils that consist of well-drained organic materials resting directly on bedrock or fragmental materials, Histosols are subdivided at the highest level on the basis of the dominant kind of peat in the subsurface tier. In the case of soils in which the control section does not extend beyond that tier or the contact between the peat and the underlying mineral soil is within it, however, the classification is based on the kind of peat in both the surface and subsurface tiers. If the peat in the specified tier or tiers is dominantly fibric, the soil is classified as a *Fibrist* at the suborder level. If hemic, it is a *Hemist,* and if sapric, it is a *Saprist.* All of the well-drained Histosols are *Folists.* Because of their importance in the water management of organic soils, a new suborder of *Limnists* for Histosols that contain layers of limnic materials has been proposed but is not now recognized in the United States Taxonomy.*

With two important exceptions, all cold Histosols are in cryic great groups. Where *sphagnum* moss peat makes up at least three-fourths of the upper 90 cm of an organic soil—that is, the entire surface tier and at least one-half of the subsurface tier—or three-quarters of the entire soil if the control section is shortened to less than 90 cm, or three-quarters of the organic portion of the soil if a mineral substratum occurs within that depth, the soil is assigned to a great group of Sphagnofibrists. This distinction is made in recognition of the unique properties of sphagnic peat and of its formation well above the normal groundwater level.

The second exception to the general rule, which applies in all suborders of Histosols, is the separation of soils that freeze deeply each year but have no perennially frozen substratum from other cold organic soils, including those with permafrost and those in areas with cool maritime climates that seldom freeze but have low temperatures throughout the year. The separated annually frozen soils, both those in the boreal belt and those with higher summer soil temperatures, are classified as Borofibrists, Borohemists, etc. These boric great groups were devised as a means of recognizing at a high taxonomic level organic soils that can be drained and used in agriculture and forestry for long periods with little loss of organic materials by decomposition, and to separate them from nonarable cold Histosols. This is a departure from the practice in other orders of the United States Taxonomy, where no comparable effort is made to separate arable from nonarable soils at the great group level; that separation is made in lower categories of the Taxonomy. Unfortunately, the boric great groups do not wholly achieve the desired result, in that they also include organic soils that thaw completely in late summer but are far too cold for any agricultural use. At the same time they introduce an illogical element into

* K. R. Everett, personal communication, December 1981.

the Taxonomy in that soils at either end of the temperature range of cold Histosols—soils that almost never freeze and soils that are perennially frozen—are grouped together, and soils in the middle of that range are in a separate group.

The number of subgroups of Histosols defined in the Taxonomy is very large and probably excessive. This is owing in part to the many and complex factors that affect the accumulation of organic materials and in part to the relatively few field correlation studies of organic soils that have been made. The classification of these soils is still considered to be tentative and imperfect. It undoubtedly will be extensively revised and simplified in the future.

A. FOLISTS

The suborder of Folists includes all well-drained Histosols. They consist of an 0 horizon that rests on bedrock or on gravel and stones with little or no fine earth in the interstices between the coarse fragments. Cold Folists occur principally in areas with maritime climates and in high mountains (Fig. 35). Many are associated with Spodosols in forested (Lewis and Lavkulich, 1972) and mountain tundra (Witty and Arnold, 1970) areas, and with Andepts, Andeptic Cryaquepts, and other Histosols in nonforested areas affected by volcanic ash (Everett, 1971). Some occur in recently deglaciated areas where the adjoining soils are Orthents and Aquepts in glacial till or sloping, poorly drained Histosols formed above the till. Others are in arctic tundra (James, 1970; Payette and Morisset, 1974) and polar desert (Tedrow and Walton, 1977) areas, where mean annual soil temperatures are lower than 0° C. The 0 horizons in all of these soils decompose at the same rate as those in associated well-drained mineral soils. They may be entirely fibrous, but the lower part of the 0 horizon, especially in forested areas where the horizon is fairly thick, may be well-decomposed with many of the characteristics of sapric peat. Except where the underlying rock is calcareous, the cold Folists are strongly acidic.

Typic Cryofolists and *Typic Borofolists* are soils in which the 0 horizon rests on fragmental material. The voids between coarse fragments may be partially filled with organic matter from the 0 horizon or from decayed roots. *Lithic Cryofolists* and *Lithic Borofolists* have 0 horizons over bedrock. Mineral soil material as much as 10 cm thick in which no diagnostic horizons used in defining other orders have developed may lie between the base of the 0 horizon and the bedrock. No distinction between soils with temperatures above and below 0° C is made at the subgroup level.

B. FIBRISTS

Typic Borofibrists are soils in which fibric peat other than *sphagnum* peat dominates the subsurface tier, that do not have shortened control sections, and

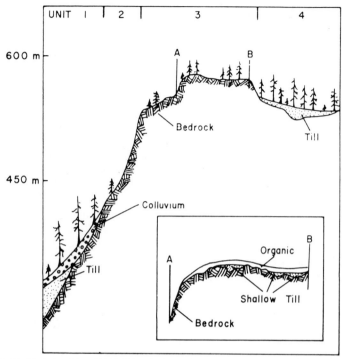

FIGURE 35. *Typical position of Lithic Folists in the rain forest of the Pacific coast. Associated with the Folists are bedrock exposures and Humic Lithic Cryorthods. (From Lewis and Lavkulich [1972] by permission of the Agricultural Institute of Canada.)*

that do not contain layers of significant thickness of hemic or sapric peat or of mineral materials within the control section. The soils, like all soils in boric great groups, freeze deeply each year but have no perennially frozen substratum. Other subgroups of Borofibrists recognize specific deviations from these criteria. *Sphagnic Borofibrists* are soils in which *sphagnum* moss peat makes up the surface tier but less than one-half of the subsurface tier. *Hemic Borofibrists* have layers of hemic peat with a total thickness of 25 cm or more in the subsurface and bottom tiers, and *Sapric Borofibrists* have layers of sapric peat with a combined thickness of 12.5 cm or more in those tiers. *Terric Borofibrists* are soils in which the lower part of the control section consists of mineral material. *Lithic Borofibrists* are soils in which the control section is shortened by bedrock. *Hydric Borofibrists* are underlain by a layer of free water, and *Limnic Borofibrists* are soils with a layer of marl, diatomaceous earth, or coprogenous earth within the control section. Soils in which either a single layer more than 5 cm thick or two or more thinner continuous layers of mineral material occur as a result of either flooding or ash falls are *Fluvaquentic Borofibrists*. These layers are particularly important in Histosols that are drained for agriculture or silviculture because of their effect on the free movement of water through the soil. Subgroups of *Sphagnic Terric Borofibrists, Hemic Terric Borofibrists,* and *Sapric Terric Borofibrists* have been established for soils in which more than one deviation from the characteristics of the Typic subgroup occurs.

The subgroups of Cryofibrists are less numerous. *Typic Cryofibrists* are like Typic Borofibrists except that they seldom freeze. The subgroup of *Pergelic Cryofibrists* includes all of the soils of the great group with mean annual temperatures of 0° C or less except for those in which bedrock is less than 25 cm below the permafrost table. *Fluvaquentic, Lithic, Sphagnic,* and *Terric Cryofibrists* are defined as are the comparable subgroups of the Borofibrists. No provision is made for Cryofibrists with layers of hemic or sapric peat because the effect of these layers on moisture retention and movement are considered to be unimportant in these cold soils.

The great group of *Sphagnofibrists* includes Histosols in which at least the upper 90 cm of the soil (or the entire organic portion of the soil if the control section is shallower than that or if a mineral substratum occurs within 90 cm) is three-fourths or more *sphagnum* moss peat. *Typic Sphagnofibrists* are defined as soils that seldom freeze; the subgroup also includes sphagnic peats of maritime areas with temperatures warmer than those of Cryofibrists. Soils of cold areas that do freeze annually but have no permafrost—that is, those with a temperature regime similar to that of the Borofibrists—are, illogically, assigned to a subgroup of *Cryic Sphagnofibrists.* Those with permafrost are in a subgroup of *Pergelic Sphagnofibrists.* Other subgroups of Sphagnofibrists are similar to those defined for the Borofibrists.

The classification of *Borohemists, Cryohemists, Borosaprists,* and *Cryosaprists* parallels in most respects that of the Borofibrists and Cryofibrists. No sphagnic subgroups are presently defined for those great groups, but it is likely that a sphagnic subgroup will be found to be needed for the Hemists.

References

Auer, V. (1930). "Peat Bogs in Southeastern Canada." Can. Dept. Mines, Geol. Surv. Mem. **162**, 1–32.

Brewer, R., Protz, R., and McKeague, J. A. (1973). Microscopy and electron microprobe analysis of some iron–manganese pans from Newfoundland. *Can. J. Soil Sci.* **53**, 349–361.

Brown, J. (1966). "Soils of the Okpilak River Region, Alaska." U. S. Army Cold Regions Res. and Eng. Lab. (CRREL) Res. Rep. **188**. Hanover, New Hampshire.

Bukhman, A. A. (1960). Characterization of the agricultural chemical properties of major peat soil groups of Karelia. *Sov. Soil Sci.,* 1233–1238.

Dachnowski-Stokes, A. P. (1941). "Peat Resources in Alaska." U. S. Dept. Agr. Tech. Bull. **769**. Washington, D. C.

Everett, K. R. (1971). Composition and genesis of the organic soils of Amchitka Island, Aleutian Islands, Alaska. *Arc. Alp. Res.* **3**, 1–16.

Farnham, R. S., and Finney, H. R. (1965). Classification and properties of organic soils. *Adv. Agron.* **17**, 115-162.

Firsova, V. P. (1967). Forest soils of the northern Transural region. *Sov. Soil Sci.,* 300-307.

Ignatenko, I. V. (1967). Waterlogged and bog soils of the East European tundra. *Sov. Soil Sci.,* 177-192.

Ingram, H. A. P. (1978). Soil layers in mires: Function and terminology. *J. Soil Sci.* **29**, 224-227.

James, P. A. (1970). The soils of the Rankin Inlet area, Keewatin, N. W. T., Canada. *Arc. Alp. Res.* **2**, 293-302.

Kovalev, V. A., and Generalova, V. A. (1967). Interaction of humic and fulvic acids with iron in peat soils. *Sov. Soil Sci.,* 1261-1268.

Lewis, T., and Lavkulich, L. M. (1972). Some Folisols in the Vancouver area, British Columbia. *Can. J. Soil Sci.* **52**, 91-98.

McKeague, J. A., Damman, A. W. H., and Heringa, P. K. (1968). Iron–manganese and other pans in some soils of Newfoundland. *Can. J. Soil Sci.* **48**, 243-253.

Page, F., DeKimpe, C. R., Bourbeau, G. A., and Rompre, M. (1980). Formations d'horizons cimentes dans le sols sableux du delta des Rivieres Manicouagan et Outardes, Quebec. *Can. J. Soil Sci.* **60**, 163-175.

Payette, S., and Morisset, P. (1974). The soils of Sleeper Islands, Hudson Bay, N. W. T., Canada. *Soil Sci.* **117**, 352-368.

Proudfoot, V. B. (1958). Problems of soil history. Podzol development at Goodland and Torr Townlands, Co. Antrim, Northern Ireland. *J. Soil Sci.* **9**, 186-198.

Retzer, J. L. (1956). Alpine soils of the Rocky Mountains. *J. Soil Sci.* **7**, 22-32.

Rieger, S., Schoephorster, D. B., and Furbush, C. E. (1979). "Exploratory Soil Survey of Alaska." U. S. Dept. Agr., Soil Cons. Serv. Washington, D. C.

Romans, J. C. C., Stevens, J. H., and Robertson, L. (1966). Alpine soils of north-east Scotland. *J. Soil Sci.* **17**, 184-199.

Silc, T., and Stanek, W. (1977). Bulk density estimation of several peats in northern Ontario using the Van Post humification scale. *Can. J. Soil Sci.* **57**, 75.

Sjörs, H. (1959). Bogs and fens in the Hudson Bay lowlands. *Arctic* **12**, 3-19.

Soil Survey Staff. (1975). "Soil Taxonomy." U. S. Dept. Agr. Handbook 436, Washington, D.C.

Tarnocai, C. (1972). Some characteristics of cryic organic soils in northern Manitoba. *Can. J. Soil Sci.* **52**, 485-496.

Tarnocai, C., and Zoltai, S. C. (1978). Soils of northern Canadian peatlands: Their characteristics and stability. *Proc. 5th N. Amer. Forest Soils Conf., Colorado State U.,* 433-448.

Tedrow, J. C. F., and Walton, G. F. (1977). Rendzina formation on Bathurst Island. *J. Soil Sci.* **28**, 519-525.

Thom, B. G. (1972). The role of spring thaw in string bog genesis. *Arctic* **25**, 236-239.

Troll, C. (1958). "Structure Soils, Solifluction, and Frost Climates of the Earth." U. S. Army Snow, Ice, and Permafrost Res. Estab. (SIPRE) Trans. 43. Hanover, New Hampshire.

Walmsley, M. E., and Lavkulich, L. M. (1975). Chemical, physical, and land-use investigations of organic terrain. *Can. J. Soil Sci.* **55**, 331-342.

Witty, J. E., and Arnold, R. W. (1970). Some Folists on Whiteface Mountain, New York. *Soil Sci. Soc. Am. Proc.* **34**, 653-657.

Zoltai, S. C., and Tarnocai, C. (1971). Properties of a wooded palsa in northern Manitoba. *Arc. Alp. Res.* **3**, 115-129.

Zoltai, S. C., and Tarnocai, C. (1975). Perennially frozen peatlands in the western Arctic and subarctic of Canada. *Can. J. Earth Sci.* **12**, 28-43.

Chapter *13*

THE TAXONOMIES
OF CANADA,
THE U.S.S.R., AND
THE FAO

The largest areas of cold soils are in Canada and the U.S.S.R. Each of these countries has its own official soil classification system that differs in important respects from the United States Taxonomy. In addition to these, the Food and Agriculture Organization (FAO) of the United Nations has developed a separate taxonomic system for a small-scale soil map of the world published by that organization. These three classification systems and the United States Taxonomy by no means exhaust the number of systems that have been proposed for the classification of soils, but it is likely that they are and will continue to be the most extensively used systems in cold areas.

The three classification systems are described briefly in this chapter and are compared with the United States Taxonomy. Each system has its advantages and disadvantages. The Canadian and U.S.S.R. classifications have been used successfully in the construction of both large and small-scale soil maps. The FAO system, which does not provide the criteria necessary for the preparation of detailed soil maps, has been used only for a world soil map at a scale of 1:5,000,000.

I. Canadian System of Soil Classification

The present Canadian System of Soil Classification (Canada Soil Survey Committee, 1978) began its development in 1955, in response to the same dissatisfaction with the 1938 United States Department of Agriculture system (also used, with some differences in nomenclature, in Canada at that time) that led to the creation of the United States Taxonomy. In the early stages of its development, there was close contact with the United States Soil survey staff, and most of the principles guiding the development of a new classification scheme are common to both taxonomies. In succeeding years, however, differences from the United States Taxonomy have emerged, at least partly because the Canadian system is designed to accommodate only the soils of Canada and is not intended to be a comprehensive taxonomy for soils of the world.

A. DIAGNOSTIC HORIZONS

Diagnostic horizons related to genetic processes have the same importance in the Canadian as in the United States Taxonomy, but only those diagnostic horizons that occur in soils of Canada are defined. Although in most cases the horizons are the same as those recognized in the United States Taxonomy, many are defined somewhat differently and have different names. Some horizons recognized as diagnostic in the United States Taxonomy—for example, the cambic and argillic horizons—are not named but are identified only by letter horizon designators.

The differences in definition of diagnostic horizons are significant in classification. The Canadian *Chernozemic A horizon,* for example, is roughly equivalent to the mollic epipedon but is required to be only 10 cm thick and is restricted to soils with drier than humid (udic) soil moisture regimes and with mean annual soil temperatures higher than 0° C. It is, therefore, largely confined to soils of the subhumid to semiarid interior plains of Canada that support grassy or mixed grass and forest vegetation. Mollic epipedons in soils of humid forested areas are recognized separately as *mull* horizons. The horizon designation, Ah, is used for Chernozemic A, mull, and other mineral surface horizons enriched with organic matter. An umbric epipedon is an Ah horizon but is excluded from the Chernozemic A horizon because of its low base saturation. The histic epipedon is not identified separately except as an 0 horizon that requires no minimum thickness (well-drained organic horizons have different letter designations) and is used in classification only as an accessory feature in the identification of poorly drained soils. There is no equivalent to the ochric epipedon in the Canadian classification system.

The *Podzolic B horizon* is essentially the same as the spodic horizon, although its chemical definition differs from that in the United States Taxonomy. Unlike the spodic horizon it is required to be at least 10 cm thick. The *Bt horizon,* although not a named diagnostic horizon, is the equivalent of the argillic horizon and is used in much the same way in classification. The *Solonetzic B* horizon corresponds to the natric horizon. The *Bm horizon* is comparable to the brown cambic horizon and the *Bg horizon* to the gleyed cambic horizon. In both cases there is no exclusion of sandy or stratified soils. The *Ae horizon* is essentially the same as the albic horizon in the United States Taxonomy. Its presence is given greater weight in classification, particularly at the subgroup level, than is the albic horizon in the United States classification. *Fragipans* and *placic horizons* are defined as they are in the United States Taxonomy.

The levels, or categories, in the Canadian classification system are basically the same as in the United States Taxonomy, but there are only five rather than six levels. There is no suborder category in the Canadian system, but classes in the great group category are defined in some orders by criteria used for separation at the suborder level in the United States Taxonomy. In other orders the differentiating criteria, except for soil temperature distinctions, are similar to those used in the United States great group level. As in the United States Taxonomy, criteria relating to moisture regimes are used at all levels above the family, but the ways in which they are used are not identical.

B. ORDERS

Nine orders are defined in the Canadian system. Four orders in the United States Taxonomy—the Aridisols, Oxisols, Ultisols, and Vertisols—are omitted because they are absent or poorly represented in Canada. The six United States orders discussed in the preceding chapters have equivalents, although far from exact in most instances, in the Canadian classification system. The Canadian orders (and their approximate United States equivalents) are the *Regosolic* order (Entisols), the *Podzolic* order (Spodosols), the *Luvisolic* order (Alfisols), the *Chernozemic* order (Mollisols), the *Brunisolic* order (Inceptisols), and the *Organic* order (Histosols). Three orders in the Canadian system group together soils that are distributed among several orders in the United States Taxonomy. These are *(a)* the *Solonetzic* order, which includes soils similar to those assigned to natric great groups of Alfisols and Mollisols except for Natraqualfs and Natraquolls; *(b)* the *Gleysolic* order, which is made up of soils that would in the United States be in aquic suborders except for those classified as Aquods or as Pergelic Cryaquepts or Pergelic Cryaquolls; and *(c)* the *Cryosolic* order, which includes all soils with permafrost tables shallower than 1 m or shallower than 2 m if the soil gives evidence of strong cryoturbation.

Although the three orders that have no United States equivalents are defined by properties that are recognized only in lower categories of the United States Taxonomy, two of them are based on characteristics acquired as a result of soil-forming processes. The designers of the Canadian classification apparently believe that the properties of soils formed under the influence of the sodium ion are unique and deserve recognition at the highest taxonomic level, that gleization as a genetic process is at least as important as any other, and that poorly drained soils have most properties in common whether or not a dark-colored horizon has developed at the surface or some clay translocation has taken place. Many earlier classification systems also reflect this view.

The major deviation from the genetic basis for classification is the Cryosolic order, which apparently was created in the belief that the presence of permafrost at shallow or moderate depth imparts properties to a soil that are more important than those that result from any genetic process. Although this idea has some validity and it is likely that the existence of permafrost should be recognized in a higher category than the subgroup (perhaps as gelic great groups) in the United States Taxonomy, an unfortunate consequence is the loss of the clear relationship between genetically similar very cold soils and soils of warmer areas that is indicated in the United States Taxonomy. This is only partially restored by intergrades to the warmer soils at the subgroup level.

The use of specific depths to the permafrost table as the sole criterion for separation of the Cryosolic order also appears to have serious deficiencies. Wildfires or clearing can have profound effects on the permafrost table; for example, a fire that removes the surface vegetation can result in recession of the permafrost table to depths greater than 1 m (or 2 m). On subsequent revegetation the permafrost table will revert to its original position. The classification of any polypedon, as a consequence, can change from the Cryosolic order to another order and back to the Cryosolic order in the space of a few decades, certainly an undesirable feature in a classification system.

C. GREAT GROUPS

Great groups in the Canadian system are named in most cases by modifying the root of the order name with a single adjective (e.g., Humic Podzol or Eutric Brunisol). There are exceptions to this rule, however. Great groups in the Chernozemic order are simply given names based on the intensity of color of the Chernozemic A horizon: that is, Brown, Dark Brown, Black, and Dark Gray, in order of increasing humidity of the soils under semiarid to subhumid climates. Great groups of the Organic order are called Fibrisols, Mesisols, Humisols—for soils with fibric, mesic (hemic), or humic (sapric) middle (subsurface) tiers—and Folisols (equivalent to the Folists). The adjective is also omitted in some great groups of the Gleysolic, Regosolic, and Solonetzic orders.

As in the United States Taxonomy, the definitive criteria for great groups are different in the different orders. The great groups of Podzolic soils, for example, are separated on the basis of the amount of organic carbon in the Podzolic B horizon and the ratio between organic carbon and pyrophosphate-extractable iron, and great groups of Cryosolic soils by whether or not evidences of strong cryoturbation exist and by whether they consist of mineral or organic material. In some cases the great groups are intergrades between the parent order and another order; for example, the Luvic Gleysol group that intergrades to the Luvisolic order and the Melanic Brunisol group that intergrades to the Chernozemic order.

D. SUBGROUPS

Subgroup names are formed by adding one or more additional adjectives to the great group name. Like the United States subgroups, the Canadian subgroups may be intergrades to other great groups or orders—for example, the Gleyed Eutric Brunisol or Podzolic Gray Luvisol subgroups—or may be extragrades as in the case of the Placic Humic Podzol or Terric Fibrisol subgroups. In several orders the presence of a thin Ae (albic) horizon, indicating greater eluviation than in the Orthic (Typic) subgroup, is the principal differentiating property. No special provision is made at either the great group or subgroup level for soils that are shallow over bedrock or for soils formed wholly or in part in pyroclastic materials.

E. EVALUATION

Overall, the Canadian classification system, although similar in concept to the United States Taxonomy, gives greater emphasis to soils such as saline and poorly drained soils that are more extensive in Canada than in the world generally and is less concerned with distinctions that are of major importance in world classifications. The soil temperature regime, except where it results in the development of permafrost, enters the system only at the family level, whereas it is a major differentiating characteristic at a high level in the United States and U.S.S.R. taxonomies. It seems unlikely that the Canadian system could be expanded to a world soil classification without at least some subordination of classes that now have greater prominence than could be warranted in a universal taxonomy.

II. Official Soil Classification of the U.S.S.R.

A. ZONAL CONCEPT

The official soil classification system of the U.S.S.R. is quite different, in both concept and form, from the United States and Canadian taxonomies. The system, although fairly new in its modern form (Rozov and Ivanova, 1967a, 1967b; Ivanova *et al.*, 1969), relies heavily on genetic principles espoused by early investigators of the nature and origin of soils in the plains of eastern

Europe. Proper classification, in this view, involves not only diagnostic traits that can be determined directly in the field or by analysis, but also the hydrological, thermal, biological, and nutrient regimes of the soil. This does not differ greatly from other taxonomies, in which at least the hydrological and thermal regimes are differentiating criteria at high levels, but the U.S.S.R. classification goes a step further in recognizing the existence of ecological-genetic or bioclimatic classes defined quantitatively in terms of the annual sums of degrees of centigrade temperature over 10°, the annual precipitation and a moisture (or leaching) coefficient, and defined qualitatively by the kind of weathering that takes place and the nature of the nutrient cycle. These classes are, in practice, expressed as geographic zones or regions with a characteristic vegetation, each of which has its own set of soil groups. The fullest expression of zonal characteristics is in soils of uplands in which soil moisture is derived entirely from precipitation with no participation by flood or groundwaters. These are the zonal soils. Actually, the definition of the zone or, in the case of zones that cover a large area, a geographic subdivision of the zone called a facies, is determined by the extent of the zonal soils within it; that is, as in other classification systems, the characteristics of the dominant soils are the reference points from which limits of classes at higher levels are determined. The profound difference between the U.S.S.R. classification and the United States and Canadian systems is that the major processes believed to be operating in the zone are used in the U.S.S.R. as the basis for classification rather than the actual properties of the soils themselves. A young loessial soil of uplands, for example, would be classified in the same group as mature soils of the region even though it had not yet developed the horizons typical of that group.

The zonal concept of soil classification has not been easily transferred to mountainous areas, where there is much greater soil variability because of aspect, steepness of slope, and other factors. To preserve the concept it has been suggested that combinations of two or more groups in each vertical zone rather than a single dominant group can be used to fully reflect the climatic factors affecting soil formation in such areas (Sokolov and Sokolova, 1962; Vishnevskaya, 1965).

B. GENETIC ORDERS

Four genetic orders based on the moisture regime are recognized within each ecological-genetic class: *(a)* the automorphic order, for soils that are affected only by atmospheric moisture; *(b)* the weakly hydromorphic-alluvial order, for soils that are wet for only short periods because of spring floods; *(c)* the semi-hydromorphic order, for soils with impeded runoff of atmospheric or floodwaters or with groundwater tables at depths between 3 and 6 m; and *(d)* the hydromorphic order, for soils subject to prolonged flooding or ponding or with water tables shallower than 3 m.

C. BIOPHYSICOCHEMICAL ORDERS

The intensity of leaching and weathering is considered to be reflected in the kind of organic acids in the soil, the minerals associated with the organic acids, and the degree of removal of soluble salts. Five qualitatively defined "biophysicochemical" orders are recognized, ranging from strongly leached acidic soils with no soluble salts or carbonates in the profile to soils with soluble salts in all horizons. Fulvic acids are dominant in the strongly leached soils and humic acids saturated with calcium and, ultimately, sodium in the drier soils. Within most ecological–genetic classes only one or two of these orders are represented, but in dry continental areas (the frozen taiga region) groups of saline soils with steppe vegetation are included as well as groups of leached forested soils.

D. SOIL GROUPS

The basic unit in the classification scheme is the soil group (or type), formally defined in part by the ecological-genetic class and in part by the genetic and biophysicochemical orders. Each soil group within an ecological-genetic class is bounded by the limits of the two orders, but it is common to have several groups within a given set of limits. The basis for separation of these groups varies. It may be the presence or absence of a "sod" upper horizon that consists of at least 5 cm of mixed mineral matter and humus, the existence of a horizon of calcium carbonate accumulation, the degree of stoniness, the character of the vegetation, or even, in the case of hydromorphic soils, the position in the landscape.

E. LOWER CATEGORIES

Below the group level are five additional levels—subgroup, genus, species, variety, and subvariety. Subgroup distinctions are based on differences in thickness or depth of certain horizons, on the presence of horizons that indicate intergrades to other groups, or on differences in the thermal or moisture regimes. These differences are believed to represent modifications of the main soil-forming process as a result of variations in zonal characteristics. Thus, each subgroup normally occurs in a particular subzone within the more broadly defined zone or facies. For example, in the temperate East European facies of the taiga-forest boreal region the strongly leached automorphic soil group, the Podzolic group, has three subgroups: (a) The Typical Podzolic soils of the central portion of the zone; (b) the Gley-Podzolic soils of the northern portion, considered to be transitional to the Gley-Tundra group to the north; and (c) the Sod-Podzolic soils of the southern portion, transitional to the Gray Forest soils of warmer regions with deciduous forests to the south.

Genera are defined by characteristics of the parent material or groundwater, by relict characteristics from former stages of soil development, by the existence of kinds of horizons that develop only in specific parent materials, or by incomplete or inconspicuous development of horizons that are characteristic of the soil group. That is, the genus category attempts to account for modifications of genetic horizons as a result of factors unrelated to the ongoing soil-forming processes in the region.

Soil species reflect the intensity of development, or the degree of expression, of fundamental processes characteristic of the soil group. The criteria for defining species vary with the group. In some soils of the Podzolic group, for example, the degree of development of a sod (A) horizon and the thickness of the podzolic (i.e., albic) and illuvial humus accumulative (spodic) horizons are important distinguishing characteristics. In other groups the content of a particular substance in one horizon or in the entire profile is of greatest importance. In most cases the criteria are defined quantitatively. Soil varieties are distinguished chiefly on the basis of soil texture, and subvarieties on lithographic and other characteristics of the parent material not considered at higher levels.

F. COMPARISON WITH UNITED STATES TAXONOMY

The differences between the U.S.S.R. and United States taxonomies are great enough so that precise correlation between taxonomic units is virtually impossible. The principal stumbling blocks are *(a)* the zonal concept in the U.S.S.R. system, which tends to presuppose soil-forming processes and, therefore, soil groups in specific geographic areas defined by climatic parameters and, indirectly, by characteristics of the native vegetation; and *(b)* the unequal importance attached to certain horizons in the two systems. The spodic horizon, for example, is the key feature in the definition of Spodosols, but a horizon with comparable properties is considered to be an aberration due primarily to properties of the parent material in the U.S.S.R. system. Most Typical Podzolic soils would be classified as Cryoboralfs in the United States Taxonomy; Typic Cryorthods would be a genus of iron–aluminum–humus Podzols in the U.S.S.R. system. This situation arises because the outstanding characteristic of Podzolic soils in the U.S.S.R. is considered to be the eluviated "podzolic" horizon, and only secondary importance is attached to the materials that accumulate in the subsurface horizon. Recognition at high levels is given to a "sod" (A) horizon in both classifications, but the horizon must be much thicker and darker in the United States than in the U.S.S.R. system (where it must be only 5 cm thick with no specific color standards). Organic soils are not given separate status in the U.S.S.R. system but are grouped with soils that have as little as 20 cm of peat accumulation at the surface. Histic

Pergelic Cryaquepts and Pergelic Cryofibrists under tundra vegetation would both be classified as Bog-Tundra soils in the U.S.S.R.; forested Histic or Histic Pergelic Cryaquepts and forested Histosols would all be included in the Upland Bog, Lowland Bog, or Alluvial Bog groups, depending on their position.

It is argued that the recognition of zones is desirable since agricultural technology consists in large part of the regulation of the hydrological, thermal, and nutrient regimes of soils, and the key to these regimes is the regional climate. That is, agricultural practices are likely to be the same on any arable soil in a given region, and the soil classification should recognize this fact (Gorshenin, 1963). There is, however, considerable opposition to the geobotanical bias of the official classification system within the U.S.S.R. Although it would be incorrect to say that all aspects of the United States Taxonomy are favored by internal critics of the U.S.S.R. classification, a basic tenet of the United States Taxonomy—that classification should be based on soil properties rather than on climatic indices or other external factors—has been repeatedly expressed (Kovda and Rozanov, 1970; Tonkonogov, 1970; Sokolov, 1976). One objection is that climatic indices consider only present-day climate. If they dominate the classification relict soil characteristics developed during earlier and different climatic regimes are given insufficient recognition despite the effect they may have on soil utilization, and no consideration is given to the importance of the time factor in soil development even under the same climate (Kovda *et al.,* 1967). The belief that only a single soil-forming process dominates in a given zone has also come under severe attack (Zonn, 1966). Even under automorphic conditions it is clear that several soil groups can form in the same zone, and, conversely, the same soils can occur in several different zones (Liverovskiy, 1969, 1977). Even some who seem to favor regionalization suggest that it should be based on soil properties rather than on presumed climatic or geobotanical correlations (Ignatenko, 1971).

G. EVALUATION

Despite the inherent defects from the United States standpoint in a soil classification system that is based, at least in part, on factors other than actual soil properties, the U.S.S.R. classification does make some broad distinctions that cannot be made readily in the United States Taxonomy. As an example, forested soils underlain by permafrost are separated on the zonal basis from soils with similar characteristics in arctic or mountain tundra areas. In the United States Taxonomy it is not always possible to make such a distinction even at the series level with existing criteria, and it becomes necessary, in some cases, to resort to unusual phase criteria in order to make satisfactory interpretations. Although it seems likely that soil temperature regimes in the two situations are not identical and that it should be possible to write definitions that will effectively distinguish perennially frozen forested soils from frozen tundra soils with similar morphology, there are presently not sufficient data to correct this deficiency in the United States Taxonomy.

The U.S.S.R. system also distinguishes, by separate zones, soils of mountainous areas from those of plains (although the line between the two is not sharply drawn) and, by the use of genetic orders involving topographic factors, soils of different positions in the landscape. In most cases the landscape position occupied by a soil is reflected in its internal characteristics, but this is not always true. It can be argued that slope, surface configuration, and topographic position are in their own right recognizable and classifiable soil properties, but except for slope as a family differentia for some poorly drained soils, it is necessary to rely on phases to make these distinctions in the U.S. system.

The concept of diagnostic horizons, although implied in some cases, is not incorporated in the U.S.S.R. classification system. This sets it apart from most other soil taxonomies that claim worldwide applicability. The principal obstacle to universal acceptance of the system, however, probably is its emphasis on regionalization. If extended over the world, it is likely that the number of zones, facies, and other geographical subdivisions would be so great that the system would become unwieldy. It would fail to satisfy a basic objective of any taxonomy, to make groupings in a way that can be easily remembered and that emphasizes major similarities and differences among all soils.

III. The FAO Soil Classification

The Food and Agriculture Organization soil classification (FAO–UNESCO, 1974) makes use of concepts and terminology from several different classification systems and, where reconciliation among them proved to be impossible, introduces some original concepts and nomenclature. The international committee that developed this scheme did not expect it to replace any of the various national soil classification systems but intended it to serve as a ''common denominator'' that would make it possible for pedologists to compare soils in their own countries with those elsewhere in the world and in that way enable them to take advantage of research and experience on similar soils in distant areas. The classification serves primarily as the basis for a small-scale world soil map prepared by the FAO.

Unlike most major taxonomic systems, which consist of a number of categories at different levels of generalization, the FAO classification has only two levels—units based on diagnostic horizons, properties of the parent materials that strongly influence soil development, or moisture regimes, and subunits defined by particular subordinate soil properties. The diagnostic horizons are essentially the same as those in the United States Taxonomy and have the same names, except that a horizon designation is appended. Thus, the mollic epipedon is called the mollic A horizon; the spodic horizon, the spodic B horizon, etc. Most horizon designations, moisture regimes used in the classification, and other soil features are also defined and named as they are in the United States Taxonomy. One exception is that the 0 horizon is reserved for well-drained, organic accumulations; saturated organic horizons are called H horizons. In all, 26 units and 103 subunits are defined in the FAO classification.

Except in the case of soils with a permafrost table within 2 m of the surface, temperature regimes are not considered. Soils with such permafrost tables in some units (but not in all that include perennially frozen soils) are assigned to *gelic* subunits; all of these would be in pergelic subgroups in the United States Taxonomy, but the United States subgroups also include soils with deeper permafrost tables. Most would be classified in the Cryosolic order of the Canadian classification system, but some with little cryoturbation and permafrost at depths greater than 1 m would be excluded.

A. CLASSIFICATION UNITS

Units in the FAO classification that are represented in cold areas are described briefly below and their relationship to classes in the United States Taxonomy is noted.

Histosols in the FAO system are much the same as Histosols in the United States Taxonomy, except that Folists are excluded and are classified according to characteristics of the mineral material below the 0 horizon. No recognition is given to the botanical origin or degree of decomposition of the peat; subunits, apart from the Gelic Histosols, are defined entirely on the basis of acidity in the 20–50 cm section.

Lithosols include all soils (except Histosols) with coherent bedrock within 10 cm of the surface of the mineral soil. Lithic Folists and the shallowest members of other lithic subgroups in the United States Taxonomy are included; no separate recognition is given to soils with bedrock deeper than 10 cm.

Fluvisols include both well- and poorly drained soils formed in recent alluvial deposits, provided that no mollic A horizon or subsurface diagnostic horizons have developed. All Cryofluvents, along with Cryaquents and Cryumbrepts in alluvial materials are included in this unit. Subunits are defined largely by chemical criteria.

Solonchaks are soils with high salinity in the upper part of the profile but with no natric horizon. They may or may not have hydromorphic properties. No provision is now made for such soils of cold areas in the United States Taxonomy, although as noted in earlier chapters, they exist in those areas.

Gleysols are soils other than those listed above that have a histic H horizon (histic epipedon), low chroma, mottling, or other properties that indicate periodic or permanent saturation in the upper 50 cm of the profile and that do not have spodic, argillic, or other diagnostic B (subsurface) horizons. Most Aquents, Aquepts, Aquolls, and aquic subgroups of Orthents and Psamments are Gleysols, but Aquods and Aqualfs are not.

Andosols are soils, other than soils with hydromorphic properties, developed in pyroclastic materials. This unit corresponds with the suborder of Andepts in the United States Taxonomy.

Arenosols are coarse-textured, well-drained soils with no surface or subsurface diagnostic horizons except for an ochric A horizon. Most of these soils are Psamments in the United States Taxonomy. Subunits are intergrades to other units on the basis of incompletely developed diagnostic horizons. There is no provision for a Gelic subunit.

Regosols are fine-textured, well-drained soils with no diagnostic horizons other than an ochric A horizon. They correspond generally with Orthents in the United States Taxonomy.

Rankers are soils with an umbric A horizon less than 25 cm thick and no other diagnostic horizon. Most would be Orthents in the United States Taxonomy, but some would classify as Umbrepts.

Rendzinas are soils with a mollic A horizon over highly calcareous material. They correspond to the Rendolls in the United States Taxonomy.

Podzols are soils with a spodic horizon. This unit is identical with the order of Spodosols in the United States Taxonomy. With two exceptions, subunits are approximately the same as suborders of Spodosols. All Podzols with placic horizons, regardless of other soil characteristics, are in a subunit of *Placic Podzols*. Soils with thin or no albic horizons and only minimal development of the spodic horizon are separated in a subunit of *Leptic Podzols;* these correspond to the Entic subgroups of Spodosols. No provision is made for Podzols with permafrost or for soils in which a spodic horizon develops in a thick albic horizon overlying an argillic horizon (Boralfic Cryorthods).

Planosols have an albic E horizon with hydromorphic properties over a slowly permeable argillic or natric horizon or a fragipan. They correspond most closely to the great group of Albaqualfs in the United States Taxonomy but also to some soils classified as Natraqualfs, Fragiaqualfs, and Cryaquolls. A few Cryaqualfs, if that great group were established, probably would also qualify as Planosols. A subunit of *Gelic Planosols* is provided for soils with a fragipan and permafrost, an unlikely combination.

Solonetz are soils with more permeable natric horizons. They include most Natriqualfs and Natriboralfs in cold areas and correspond generally with the Canadian Solonetzic order.

Greyzems are soils of moderately dry areas with a mollic A horizon and evidence of leaching in the form of bleached coatings on ped surfaces. They may or may not have an argillic horizon. In cold areas most would be included with the great groups of Cryoborolls or Cryaquolls. A subunit of *Gleyic Greyzems* is provided for soils with hydromorphic properties and an argillic horizon.

Chernozems have a dark mollic A horizon and an accumulation of calcium carbonate or gypsum in the subsoil. They may or may not have an argillic horizon. In cold areas these soils would be classified as Cryoborolls in the United States Taxonomy. In contrast with the large number of subgroups of Cryoborolls in the United States system, only four subunits of Chernozems are defined in the FAO classification. Soils with an argillic B horizon are *Luvic Chernozems;* those with tonguing of the A horizon into a cambic B horizon or into the C horizon are *Glossic Chernozems;* those with no argillic horizon but with sufficient accumulation of calcium carbonate or calcium sulfate to constitute a calcic or gypsic horizon are *Calcic Chernozems;* and all others are *Haplic Chernozems.* No separate provision is made for Chernozems with permafrost.

Kastanozems are similar to Chernozems but occur under drier climates and have mollic A horizons with relatively light colors. In cold areas they, also, classify as Cryoborolls in the United States Taxonomy.

Phaeozems are more strongly leached than Chernozems but have mollic A horizons. Some have argillic horizons and some hydromorphic properties in addition to the argillic horizon. They are mostly Albolls, Aquolls, or Argiaquic, Boralfic, or Vertic Cryoborolls in the United States Taxonomy.

Podzoluvisols have argillic horizons that are deeply penetrated by tongues of albic material or that are strongly nodulated. Most of these soils in cold regions would be classified as Glossic Cryoboralfs in the United States Taxonomy.

Luvisols are soils with no mollic A horizon and an argillic B horizon. They would be classified as Alfisols in the United States Taxonomy.

Cambisols are soils with a cambic B horizon with no evidence of gleying or an umbric A horizon more than 25 cm thick and with no other horizons that are diagnostic for other orders. They correspond with Ochrepts and Umbrepts in the United States Taxonomy. A *gelic* subunit is provided.

B. EVALUATION

The principal value of the FAO classification system and the world soil map based on it is that it can be used as a means of comparison of soils of different countries by pedologists familiar only with their own national classification system. Such comparisons can be only general and imprecise, however. The attempt to develop a taxonomic system that would have universal acceptance is laudable, but much more needs to be done before this system could substitute adequately for any of the national taxonomies now in use. A major deficiency is its failure to make use of soil temperature regimes as differentiating criteria except in the case of soils with fairly shallow permafrost tables in some units. The weight given to properties like base saturation and accumulations of carbonates or gypsum in the definitions of subunits seems, at least in some cases, to be excessive. Conversely, in other parts of the scheme, not enough weight is given to properties related to genesis, such as the argillic horizon. These aspects of the FAO system have the effect of making more difficult the conversion from this to other classification systems.

Approximate equivalents of subgroups in the Canadian system and subunits in the FAO system to subgroups in the United States Taxonomy are indicated in Table I.

TABLE I.

Approximate Equivalents of Subgroups in the United States Soil Taxonomy in the Canadian and FAO Soil Classification Systems.[a]

United States Taxonomy (subgroups)	Canadian system (subgroups)	FAO system (subunits)
ENTISOLS		
Aquents		
Typic Cryaquents	Rego Gleysol	Eutric (Dystric, Calcaric) Gleysols
Andaqueptic Cryaquents	Rego Gleysol	No provision[b]
Fluvents		
Typic Cryofluvents	Cumulic Regososl	Eutric (Dystric, Calcaric) Fluvisols
Andeptic Cryofluvents	No provision	No provision
Aquic Cryofluvents	Gleyed Cumulic Regosol	Eutric (Dystric, Calcaric) Fluvisols
Mollic Cryofluvents	Cumulic Humic Regosol	Eutric (Calcaric) Fluvisols
Orthents		
Typic Cryorthents	Orthic Regosol	Eutric (Dystric, Calcaric) Regosols
Alfic Andeptic Cryorthents	Orthic Regosol	No provision
Andeptic Cryorthents	No provision	No provision
Aquic Cryorthents	Gleyed Regosol	Eutric (Dystric, Calcaric) Gleysols
Lithic Cryorthents	No provision	Lithosols (in part)
Pergelic Cryorthents	Regosolic Turbic (Static) Cryosol	Gelic Regosols
Psamments		
Typic Cryopsamments	Orthic Regosol	Albic Arenosols
Alfic Cryopsamments	Orthic Regosol	Luvic Arenosols
Aquic Cryopsamments	Gleyed Regosol	Eutric (Dystric, Calcaric) Gleysols
Lithic Cryopsamments	No provision	Lithosols (in part)

Table I. (Continued)

United States Taxonomy (subgroups)	Canadian system (subgroups)	FAO system (subunits)
Pergelic Cryopsam- ments	Regosolic Static Cryosol	No provision
Spodic Cryopsamments	Orthic Humo-Ferric Podzol	Leptic Podzols
SPODOSOLS		
Aquods		
Typic Cryaquods	Orthic Humic Podzol Ortstein Humic Podzol	Gleyic Podzols
Lithic Cryaquods	No provision	Lithosols, Gleyic Podzols
Pergelic Cryaquods	No provision	No provision
Pergelic Sideric Crya- quods	No provision	No provision
Sideric Cryaquods	Gleyed Ferro–Humic Podzol	Gleyic Podzols
Cryic Fragiaquods	Fragic Humic Podzol	No provision
Cryic Placaquods	Placic Humic Podzol	Placic (Gleyic) Podzols
Placic Haplaquods	Placic Humic Podzol	Placic (Gleyic) Podzols
Ferrods		
No subgroups	Orthic Humo–Ferric Podzol	Ferric Podzols
Humods		
Typic Cryohumods	Orthic Humic Podzol	Humic Podzols
Haplic Cryohumods	Orthic Humic Podzol	Humic Podzols
Lithic Cryohumods	No provision	Lithosols (in part)
Pergelic Cryohumods	No provision	No provision
Cryic Placohumods	Placic Humic Podzol	Placic Podzols
Orthods		
Typic Cryorthods	Orthic Ferro-Humic Podzol	Orthic Podzols
Boralfic Cryorthods	Luvisolic Ferro-Humic Podzol Podzolic Gray Brown Luvisol	No provision
Entic Cryorthods	Orthic Humo-Ferric Podzol	Leptic Podzols
Humic Cryorthods	Orthic Ferro-Humic Podzol	Orthic Podzols
Humic Lithic Cryor- thods	Orthic Ferro-Humic Podzol	Orthic Podzols
Lithic Cryorthods	No provision	Lithosols (in part)
Pergelic Cryorthods	No provision	No provision
Cryic Fragiorthods	Fragic Ferro-Humic Podzol	No provision
Cryic Placorthods[c]	Placic Ferro-Humic Podzol	Placic Podzols
ALFISOLS		
Aqualfs		
Typic Cryaqualfs[c]	Orthic Luvic Gleysol	Gleyic Luvisols
Glossic Cryaqualfs[c]	Orthic Luvic Gleysol	Gleyic Luvisols
Paleo Cryaqualfs[c]	No provision	Gleyic Luvisols
Pergelic Cryaqualfs[c]	No provision	No provision
Cryic Natraqualfs[c]	Humic Luvic Gleysol Gleyed Dark Brown (Black) Solonetz	Gleyic Solonetz
Pergelic Natraqualfs[c]	No provision	No provision

(continued)

Table I. (Continued)

United States Taxonomy (subgroups)	Canadian system (subgroups)	FAO system (subunits)
Boralfs		
Typic Cryoboralfs	Orthic Gray Brown Luvisol	Albic Luvisols
Andeptic Cryoboralfs	No provision	No provision
Aquic Cryoboralfs	Gleyed Gray Brown Luvisol	Gleyic Luvisols
Glossic Cryoboralfs	No provision	Eutric (Dystric) Podzoluvisols
Lithic Cryoboralfs	No provision	No provision
Lithic Mollic Cryoboralfs	Orthic Gray Brown Luvisol	No provision
Mollic Cryoboralfs	Orthic Gray Brown Luvisol	Albic Luvisols
Pergelic Cryoboralfs	No provision	No provision
Psammentic Cryoboralfs	Orthic Gray Brown Luvisol	No provision
Cryic Natriboralfs[c]	Dark Brown (Black) Solodized Solonetz	Mollic (Orthic) Solonetz
Pergelic Natriboralfs[c]	No provision	No provision
MOLLISOLS		
Albolls		
Typic Cryalbolls[c]	Solonetzic Gray Luvisol	Eutric (Dystric) Planosols
Aquolls		
Typic Cryaquolls	Orthic Humic Gleysol	Gleyic Greyzems
Argic Cryaquolls	Humic Luvic Gleysol	Gleyic Phaeozems
Calcic Cryaquolls	Orthic (Rego) Humic Gleysol	Gleyic Greyzems
Cumulic Cryaquolls	Orthic Humic Gleysol	Gleyic Greyzems
Histic Cryaquolls	Orthic Humic Gleysol	Gleyic Greyzems
Pergelic Cryaquolls	Gleysolic Turbic (Static) Cryosol	No provision
Thapto–Histic Cryaquolls	No provision	No provision
Borolls		
Typic Cryoborolls	Orthic Black	Haplic Chernozems
Abruptic Cryoborolls	Solonetzic Black	Mollic Planosols
Andeptic Cryoborolls	No provision	No provision
Aquic Cryoborolls	Gleyed Rego Black	Gleyic Greyzems
Argiaquic Cryoborolls	Gleyed Eluviated Black	Gleyic Phaeozems
Argic Cryoborolls	Eluviated Black	Luvic Chernozems
Argic Lithic Cryoborolls	Eluviated Black	Luvic Chernozems
Argic Pachic Cryoborolls	Orthic Dark Gray	Luvic Chernozems
Argic Vertic Cryoborolls	No provision	No provision
Boralfic Cryoborolls	Eluviated Black	No provision
Boralfic Lithic Cryoborolls	Eluviated Black	No provision
Calcic Cryoborolls	Calcareous Black	Calcic Chernozems

Table I. (Continued)

United States Taxonomy (subgroups)	Canadian system (subgroups)	FAO system (subunits)
Calcic Pachic Cryoborolls	Calcareous Black	Calcic Chernozems
Cumulic Cryoborolls	Rego Black	Haplic Chernozems
Duric Cryoborolls	No provision	No provision
Lithic Cryoborolls	No provision	Lithosols (in part)
Lithic Ruptic–Argic Cryoborolls	No provision	Lithosols (in part)
Lithic Ruptic–Entic Cryoborolls	Orthic Turbic Cryosol	Lithosols (in part)
Natric Cryoborolls	Black Solonetz	Mollic Solonetz
	Black Solodized Solonetz	
Pachic Cryoborolls	Rego Black	Haplic Chernozems
Pergelic Cryoborolls	Orthic Static Cryosol	No provision
Pergelic Ruptic–Entic Cryoborolls[c]	Orthic Turbic Cryosol	No provision
Vertic Cryoborolls	Rego Black	Haplic Chernozems
Rendolls		
Cryic Rendolls	Rego Black	Rendzinas
Cryic Lithic Rendolls	No provision	Lithosols; Rendzinas
INCEPTISOLS		
Andepts		
Typic Cryandepts	No provision	Humic (Mollic) Andosols
Dystric Cryandepts	No provision	Humic (Mollic) Andosols
Dystric Lithic Cryandepts	No provision	Lithosols (in part)
Entic Cryandepts	No provision	Vitric (Ochric) Andosols
Lithic Cryandepts	No provision	Lithosols (in part)
Aquepts		
Typic Cryaquepts	Rego (Orthic) Gleysol	Eutric (Dystric, Calcaric) Gleysols
Aeric Cryaquepts	Fera Gleysol	Eutric (Dystric, Calcaric) Gleysols
Aeric Humic Cryaquepts	Fera Humic Gleysol	Humic Gleysols
Andic Cryaquepts	No provision	No provision
Halic Cryaquepts[c]	Orthic Gleysol	Gleyic Solonchaks
Halic Pergelic Cryaquepts[c]	Gleysolic Static (Turbic) Cryosol	Gelic Gleysols
Histic Cryaquepts	Rego (Orthic) Gleysol	Eutric (Dystric, Calcaric) Gleysols
Histic Lithic Cryaquepts	Rego (Orthic) Gleysol	Lithosols (in part)
Histic Pergelic Cryaquepts	Gleysolic Static (Turbic) Cryosol	Gelic Gleysols
Humic Cryaquepts	Rego (Orthic) Humic Gleysol	Humic Gleysols

(*continued*)

Table I. (Continued)

United States Taxonomy (subgroups)	Canadian system (subgroups)	FAO system (subunits)
Humic Pergelic Cryaquepts	Gleysolic Static (Turbic) Cryosol	Gelic Gleysols
Lithic Cryaquepts[c]	No provision	Lithosols (in part)
Pergelic Cryaquepts	Gleysolic Static (Turbic) Cryosol	Gelic Gleysols
Pergelic Ruptic–Histic Cryaquepts	Gleysolic Turbic Cryosol	Gelic Gleysols
Ochrepts		
Typic Cryochrepts	Orthic (Eluviated) Eutric Brunisol	Eutric (Calcic) Cambisols
Alfic Cryochrepts	Eluviated Eutric Brunisol	Eutric (Calcic) Cambisols
Andic Cryochrepts	No provision	No provision
Aquic Cryochrepts	Gleyed Eutric (Dystric) Brunisol	Gleyic Cambisols
Dystric Cryochrepts	Orthic (Eluviated) Dystric Brunisol	Dystric Cambisols
Lithic Cryochrepts	No provision	Lithosols (in part)
Pergelic Cryochrepts	Brunisolic Static (Turbic) Brunisol	Gelic Cambisols
Umbrepts		
Typic Cryumbrepts	Orthic (Eluviated) Sombric Brunisol	Dystric Cambisols
Andic Cryumbrepts	No provision	No provision
Aquic Cryumbrepts	Gleyed (Eluviated) Sombric Brunisol	Gleyic Cambisols
Entic Cryumbrepts	Orthic Humic Regosol	Humic Cambisols
Lithic Cryumbrepts	No provision	Lithosols (in part)
Lithic Ruptic–Entic Cryumbrepts	No provision	Lithosols (in part)
Pergelic Cryumbrepts	Brunisolic Static Cryosol	Gelic Cambisols
Pergelic Ruptic–Entic Cryumbrepts	Brunisolic Turbic Cryosol	Gelic Cambisols
Ruptic–Lithic Cryumbrepts	No provision	Lithosols (in part)
HISTOSOLS		
Fibrists		All organic soils except
Typic Borofibrists	Typic Fibrisol	Lithosols and Gelic
Fluvaquentic Borofibrists	Cumulo Fibrisol	Histosols are either
		Dystric Histosols (pH in
Hemic Borofibrists	Mesic Fibrisol	water less than 5.5 in some
Hemic Terric Borofibrists	Terric Mesic Fibrisol	layer between 20 and 50
		cm from the surface) or
Hydric Borofibrists	Hydric Fibrisol	Eutric Histosols (all
Limnic Borofibrists	Limno Fibrisol	others).
Lithic Borofibrists	No provision	
Sapric Borofibrists	Humic Fibrisol	

Table I. (Continued)

United States Taxonomy (subgroups)	Canadian system (subgroups)	FAO system (subunits)
Sapric Terric Borofibrists	Terric Humic Fibrisol	
Sphagnic Borofibrists	Typic Fibrisol	
Sphagnic Terric Borofibrists	Terric Fibrisol	
Terric Borofibrists	Terric Fibrisol	
Typic Cryofibrists	Typic Fibrisol	
Fluvaquentic Cryofibrists	Cumulo Fibrisol	
Lithic Cryofibrists	No provision	
Pergelic Cryofibrists	Fibric Organic Cryosol	Gelic Histosols
	Terric Fibric Organic Cryosol	
	Glacic Organic Cryosol	
Sphagnic Cryofibrists	Typic Fibrisol	
Terric Cryofibrists	Terric Fibrisol	
Typic Sphagnofibrists	Typic Fibrisol	
Cryic Sphagnofibrists	Typic Fibrisol	
Pergelic Sphagnofibrists	Fibric Organic Cryosol	Gelic Histosols
	Terric Fibric Organic Cryosol	
	Glacic Organic Cryosol	
Folists		
Typic Borofolists	Typic Folisol	Lithosols
Lithic Borofolists	Typic Folisol	Lithosols
Typic Cryofolists	Typic Folisol	Lithosols
Lithic Cryofolists	Typic Folisol	Lithosols
Hemists		
Typic Borohemists	Typic Mesisol	
Fibric Borohemists	Fibric Mesisol	
Fibric Terric Borohemists	Terric Fibric Mesisol	
Fluvaquentic Borohemists	Cumulo Mesisol	
Hydric Borohemists	Hydric Mesisol	
Limnic Borohemists	Limno Mesisol	
Lithic Borohemists	No provision	
Sapric Borohemists	Humic Mesisol	
Sapric Terric Borohemists	Terric Humic Mesisol	
Terric Borohemists	Terric Mesisol	
Typic Cryohemists	Typic Mesisol	
Fluvaquentic Cryohemists	Cumulo Mesisol	
Lithic Cryohemists	No provision	
Pergelic Cryohemists	Mesic Organic Cryosol	Gelic Histosols
	Terric Mesic Organic Cryosol	

(continued)

Table I. (Continued)

United States Taxonomy (subgroups)	Canadian system (subgroups)	FAO system (subunits)
	Glacic Organic Cryosol	
Terric Cryohemists	Terric Mesisol	
Saprists		
Typic Borosaprists	Typic Humisol	
Fibric Borosaprists	Fibric Humisol	
Fibric Terric Borosa-prists	Terric Fibric Humisol	
Fluvaquentic Borosa-prists	Cumulo Humisol	
Hemic Borosaprists	Mesic Humisol	
Hemic Terric Borosa-prists	Terric Mesic Humisol	
Limnic Borosaprists	Limno Humisol	
Lithic Borosaprists	No provision	
Terric Borosaprists	Terric Humisol	
Typic Cryosaprists	Typic Humisol	
Fluvaquentic Cryosa-prists	Cumulo Humisol	
Lithic Cryosaprists	No provision	
Pergelic Cryosaprists	Humic Organic Cryosol	Gelic Histosols
	Terric Humic Organic Cryosol	
	Glacic Organic Cryosol	
Terric Cryosaprists	Terric Humisol	

[a] Classification of soils in the three systems seldom correspond exactly. Only those subgroups in the Canadian system and subunits in the FAO system that most closely correspond with the subgroup in the United States Taxonomy are listed here. In most cases some soils of the United States subgroup are properly classified in closely related subgroups or subunits of the other systems. This is especially true for Mollisols, which may be classified in Brown, Dark Brown, or Dark Gray as well as Black great groups in the Canadian system, and in subunits of Greyzems or Kastanozems as well as Chernozems in the FAO system.

[b] "No provision" means that no specific recognition is given to the characteristic that distinguishes the subgroup in the United States Taxonomy. The soils would be classified according to characteristics other than those that define the United States subgroup.

[c] This subgroup is not yet officially recognized in the United States Taxonomy.

References

Canada Soil Survey Committee, Subcommittee on Soil Classification. (1978). "The Canadian System of Soil Classification." Can. Dept. Agr. Pub. 1646. Supply and Services Canada, Ottawa.

Food and Agriculture Organization (FAO–UNESCO) (1974). "Soil Map of the World, Vol. 1, Legend." UNESCO, Paris.

Gorshenin, K. P. (1963). Basic principles in the classification of Siberian soils. *Sov. Soil Sci.*, 15–19.

Ignatenko, I. V. (1971). Soils in the Kara River basin and their zonal position. *Sov. Soil Sci.* **3**, 15-28.

Ivanova, Ye. N., Lobova, Ye. V., Nogina, N. A., Rozov, N. N., Fridland, V. M., and Shuvalev, S. A. (1969). Present status of the doctrine of soil genesis in the U.S.S.R. *Sov. Soil Sci.* **1**, 265-277.

Kovda, V. A., and Rozanov, B. G. (1970). International FAO/UNESCO world soil map (scale 1:5,000,000) project. *Sov. Soil Sci.* **2**, 28-40.

Kovda, V. A., Lobova, Ye. V., and Rozanov, B. G. (1967). Classification of the world's soils—general considerations. *Sov. Soil Sci.*, 427-441, 851-863.

Liverovskiy, Yu. A. (1969). Some unresolved problems in classification and systemization of U.S.S.R. soils. *Sov. Soil Sci.* **1**, 106-116.

Liverovskiy, Yu. A. (1977). Soil classification problems. *Sov. Soil Sci.* **9**, 428-436.

Rozov, N. N., and Ivanova, Ye. N. (1967). Classification of the soils of the U.S.S.R. (Principles and a systematic list of soil groups). *Sov. Soil Sci.*, 147-155. (a)

Rozov, N. N., and Ivanova, Ye. N. (1967). Classification of the soils of the U.S.S.R. (Genetic description and identification of the principal subdivisions). *Sov. Soil Sci.*, 288-299. (b)

Sokolov, I. A. (1967). Major geographic–genetic soil concepts and terms. *Sov. Soil Sci.* **8**, 711-723.

Sokolov, I. A., and Sokolova, T. A. (1962). Zonal soil groups in permafrost regions. *Sov. Soil. Sci.*, 1130-1136.

Tonkonogov, V. D. (1970). Genesis of manganese–iron new formations in sandy Podzols. *Sov. Soil Sci.* **2**, 159-167.

Vishnevskaya, I. V. (1965). Mountain-taiga soils of the high taiga in eastern Tuva. *Sov. Soil Sci.*, 903-910.

Zonn, S. V. (1966). Development of Brown Earths, Pseudopodzols, and Podzols. *Sov. Soil Sci.*, 751-758.

INDEX

230   *Index*